Branding and Promotion of Indian Plantation Commodities

The Author

Dr. S. John Mano Raj holds Ph.D in Rural Marketing from M.S. University, Tirunelveli, Tamil Nadu, India, MBA in Marketing and also holds an under graduation in Engineering. Currently serving as Associate Professor (Marketing) and Chairperson for the PG Programme at Indian Institute of Plantation Management, Bengaluru, an Autonomous Organization of Ministry of Commerce and Industry, Govt. of India. He has about 24 years of Post Graduate teaching experience with a blend of administrative and industry experience through organizing and conducting training, MDPs, research and consultancy works. Designed and conducted several training programmes for the stakeholders in the plantation sector and Commodity Boards. Designed and coordinated Management Development Programmes (MDPs) for the Corporate in Plantation and Agribusiness sector. Carried out research and consultancy projects funded by various agencies such as Ministry of Commerce and Industry, Spices Board , Ministry of Consumer Affairs and ICSSR, Ministry of HRD, New Delhi. Published several research papers in refereed journals, few chapters in books and participated in several international conferences and presented papers. Undergone various FDPs including IIMK and Anna University. He has received the "Best Teacher Award in Marketing Management" from national Leadership Awards in 2015.

ISBN 9789387057562 (International Edition)

Published by : **Scholars World**
A Division of
Astral International Pvt. Ltd.
– ISO 9001:2015 Certified Company –
4736/23, Ansari Road, Darya Ganj
New Delhi-110 002
Ph. 011-4354 9197, 2327 8134
E-mail: info@astralint.com
Website: www.astralint.com

Branding and Promotion of Indian Plantation Commodities

Dr. S. John Mano Raj

2018

Scholars World

A Division of

Astral International Pvt. Ltd.

New Delhi – 110 002

Acknowledgement

I take this opportunity to express my sincere gratitude to all those who directly and indirectly contributed for the successful completion of the project and the report. First and foremost, I owe a great debt of gratitude to **Prof. Pranab Banerji**, Project Director, and **Prof.Suresh Mishra**, Coordinator, Consultancy project sponsored by Ministry of Consumer Affairs on Consumer Protection and Consumer Welfare, New Delhi and the evaluation committee for considering the project for funding.

Prof. Dr. V. G. Dhanakumar, Director of Indian Institute of Plantation Management, Bangalore has always been a source of inspiration in all my endeavors and my sincere thanks to him for all his moral support.

I am thankful to **Shri. Jawaid Akhtar, IAS** Chairman Coffee Board of India for facilitating a coffee stakeholders meet at Coffee Board office Bangalore. My special thanks to **Shri. Soundararajan**, Director (Development), Tea Board of India for arranging interaction with the officials of Tea Board at Kolkata.

The cooperation shown by all the respondents in carrying out the field survey both online and in person was commendable. I wish to extend my sincere thanks to all of them for sparing their valuable time and taking part in the survey.

Ms. Ramya Vinodh, Research Associate need a special thanks who assisted me in the project and all the field investigators for carrying out the field survey across Karnataka state. My thanks and appreciation to the staff of institute contributed directly or indirectly for the completion of the work.

Dr. S. John Mano Raj
Project Director
Associate Professor (Marketing)
Indian Institute of Plantation Management

Foreword

I am very much delighted to know that a research report based on comprehensive work on consumer welfare in general and plantation commodities in particular. The plantation sector which is undergoing several structural infirmities and one among the major forces responsible for this is the demand side shift.

The report reflects on the variables such as awareness level of consumers, brand evaluation criteria, attitude towards foreign brands, consumer ethnocentrism and perceived country image which play a significant role in promoting agri-commodity on 'country of origin labelling'. The influence of these variables is high, however they are les understood for Indian consumer in general and coffee/ tea consumer in particular.

Looking through the outcome of the study and the suggestions offered, I am convinced that it will be of great relevance to all the stakeholders in the coffee and tea sector. The planters can ensure standardization of the produce at their level, which is the basic requirement for branding effort, which fetches them a remunerative price. Creating a logo and the making labelling mandatory to the processors and marketers promoting the same would bring in more awareness and trust among the consumers and above all a safe consumption of coffee/tea.

In a way, such study is unique and I appreciate the effort made by Dr.S.John Mano Raj to come out with this work which very much need of the sector. My sincere gratitude to the sponsoring organization for funding the project.

Dr. V. G. Dhanakumar
Director
Indian Institute of Plantation Management
Bangalore

Executive Summary

Globalization of markets presents considerable challenges and opportunities for domestic and international marketers. Consumers around the world have been increasingly exposed to foreign products, giving them more buying choices. The demand supply situations for the selected plantation commodities such as tea and coffee in the Asian continent have undergone a rapid transformation due to the growth of the world economy and lowering of trade barriers resulting into surge in import. To give a barrier to such a situation, it is high time to create awareness on the positive characteristics of Indian Coffee/ tea and initiate appropriate branding and promotional strategies especially on country of origin labelling.

The literature pertaining to the research issues has been reviewed rigorously and identified the following variables very much relevant for promoting the products on 'country of origin labeling' (COOL):

- Awareness level of Indian consumers on coffee / tea producing countries, quality and brands.
- The Criteria considered by the consumers in evaluation of coffee/tea brands and the relative importance given to the country of origin.
- Consumers' attitude towards foreign brands
- Consumer Ethnocentrism i.e. consumers' affinity and tendency to buy domestic brands and
- The consumers' perceived image on India as a country.

A suitable conceptual model incorporating the above mentioned variables has been developed based on the existing literature available and suiting to the current study. Accordingly, for the successful promotion on COOL, the consumers' awareness level should be more, relative importance given to country of origin in the brand evaluation should be higher, their attitude towards foreign brands should be negative, and consumer ethnocentric tendency should be higher and finally, they should perceive a positive image about the country. The review of literature revealed that these variables have an influence on the purchase intention. But, no such organized study has been carried out to measure these variables among Indian consumers of coffee/tea.

With this as a background, rationale and conceptual model, a field survey has been carried out across the Karnataka state with the following objectives:

- To explore the level of awareness among the consumers of coffee and tea regarding the places where these commodities are produced and the quality.
- To determine the extent of 'country of origin' effect compared to other aspects upon product evaluation and decision making.
- To measure the attitude of consumers towards foreign brands of coffee and tea
- To assess the Indian consumers ethnocentrism pertaining to the purchase of tea and coffee.
- To examine the perceived image of India as a country among the domestic consumers.

Survey has been conducted among different consumer category across the state through a field survey and also online survey using a structured questionnaire. The questionnaire contained sets of questions pertaining to the research variables.

Samples were chosen from entire Karnataka state and the sampling unit is individual consumer.

The four revenue divisions were identified such as Bengaluru, Mysure, Bealgavi and Gulbarga. Samples size of 375, were drawn from each revenue divisions representing different categories such as: Student , Home maker, Agri. and allied activities, Salaried/ Entrepreneurs/ Self Business, Others. The statistical analysis such as ANOVA, t-test, f-test, factor analysis were deployed to bring inferences.

The study revealed that the level of consumers' awareness on the places where coffee/tea grown globally and their brands is very poor. Moreover, the awareness on quality of coffee/tea produced in India is also very poor. Consumers consider many attributes for brand evaluation and highest being the 'taste' and lowest is 'country of origin'. The attitude towards foreign brands is though favourable, it is not much significant. The consumer ethnocentrism and perceived country image is at the moderate level. The inferences of the study brought in clarity that both coffee and tea can be promoted on the country of origin labelling to reach the desired objective of protection of consumer and the domestic market. A three pronged strategy is being suggested for the successful promotion, viz. consumer education, regulations and mandatory labeling and finally, branding and brand promotion.

Awareness has to be created with respect to the places in India where coffee and tea is grown, the special characteristics of coffee / tea grown at different elevations and agro-climatical conditions, soil, traceability, the plant, harvesting method and also the preparation method which helps them to go for a responsible consumption. Suggested to have the visual appeal such as colour, photography, artistic renderings, certification seals, company logos, brand India coffee/tea logo maps, etc., in the package to disseminate information and educate the consumer. It is also suggested to use technology such as QR codes for interacting with consumer. Consumer education through proper agencies such as SHGs and other voluntary organizations would be beneficial.

Making origin labelling as mandatory for coffee/tea produced, processed, packed and marketed within the country would be the second suggested strategy through which the consumers' interest and domestic market can be protected. Granting the license and permission to the coffee/tea marketers who meet the desired requirement, not just help them to promote their brand, it also puts pressure on the planters, processors and marketers to strictly adhere to the standards set. Thus, it ensures that the consumer is not exploited with inferior quality products.

Since labelling helps in creating a strong distinction in the market place and acts as a vehicle for marketing, it is suggested to deploy suitable promotional measure to leverage the commercial benefits of the same. Considering the long value chain in coffee/tea and the value contributed by different market functionaries, suggestion is offered to do the branding efforts with a holistic approach. It ensure that all the market functionaries share the responsibilities in brand promotion and reap the benefit of the same. Planters ensure that there is a standardization of the produce at their level, which is the basic requirement for branding effort and they get a remunerative price. Creating a logo and the vocabulary and making it mandatory to the processors and marketers would bring in more trust, confidence and comfort among the intermediaries and gives them strength to actively participate in trade. Brand promotion to be initiated aiming at the final consumer to establish brand image and loyalty among the consumers. This will create consumer demand and also acts as a strong barrier for any other inferior brands of coffee/tea entering into the Indian market.

Suggestions are also offered to initiate suitable generic promotional strategies to bring in awareness among the consumers on the importance of 'origin' among the consumers. Any country trying to build image globally need to build strong image in the home country. Suggested to put efforts to strengthen the people's confidence on political system, stabilize the economy, technological advancement, etc. to build the image of the country.

The study had few limitations such as confining to Karnataka state, selected plantation commodities, etc. Suggestions were given for future research studies such as similar studies across the country, covering different categories of products. More importantly, such studies on promotion based on 'country of origin labelling' found to be very much relevant to international marketing strategy. Though the inferences and the outcome of the study act as a base for international promotion, the same cannot be replicated for export market of coffee and tea. Therefore, an in-depth study on the feasibility of branding on 'country of origin labelling' need to be carried out separately for coffee and tea among the consumers of coffee and tea in the emerging importing countries.

Dr. S. John Mano Raj
Project Director
Associate Professor (Marketing)
Indian Institute of Plantation Management

Contents

Introduction

1.1. Background

Plantation sector is a part of agriculture sector and plantation crops are export oriented commodities, contributing to foreign exchange earnings. The major plantation crops in India are tea, coffee, rubber, spices, coconut and cashew nut. While the development of tea, coffee, spices and rubber come under the Ministry of Commerce, coconut and cashew nut are under the Ministry of Agriculture. Plantation Sector contributes to the employment both directly and indirectly through its forward and backward linkages. According to the estimates of Ministry of commerce more than two million people are engaged in plantation sector directly and another six million are indirectly engaged in plantation sector. In India plantation crops are mainly grown in Kerala, Karnataka, Tamil Nadu, West Bengal and North Eastern states of India. Being export oriented crops, with the growing economic integration among the countries of the world in recent years, the plantation sector of India is exposed to heightened international competition. The Indian plantation industry faces a stiff competition from emerging producing countries such as Indonesia, Vietnam, Kenya, etc.

These commodities have been losing its vital share in the international markets over the years. Increasing global consumer awareness of products from different countries, increasing quality expectations, trade regulations etc. are some of the major factors responsible for the decline in the market share. Moreover, the demand supply situations in the Asian continent have undergone a rapid transformation due to the growth of the world economy and lowering of trade barriers. The formation of regional trading blocks like ASEAN – India Free Trade Agreement (AIFTA), Bangkok Agreement, South Asia Free Trade Agreement, etc. has given rise to powerful associations with strong bargaining power and significantly influences the demand and supply factors in the global market. India's tariff rates for the selected plantation products such as tea and coffee are quite high by international standards and the AIFTA envisages that the tariff rates in these commodities will be brought down in a phased manner during 2010 – 19. The tariff reduction lead to a significant increase in India's imports from the ASEAN countries. The surge in imports will have an adverse effect on Indian farmers and also the customers. The domestic market which is price sensitive will be lured by the inferior quality tea produced from Vietnam and coffee produced by Indonesia. This causes a serious threat on the domestic market for the plantation products produced in India.

Therefore, in an open economy, competitiveness of a product is not only important from export point of view, it is equally important to protect and survive in the domestic market. Moreover, in recent years the domestic consumption for the products has registered a positive growth rate which needs attention. The shift in increasing consumption in the producing countries demands various promotional schemes to leverage favorably to the Indian economy. Considering these dynamics in the export and domestic market, and the structural changes, it is essential to be creative in marketing these products. It is a high time for India to be innovative in promotion of plantation products so as to protect the market. Creating brand, building it and bring equity could be one such creative strategy that not just acts as an entry barrier, branding is also an important means to reduce the price uncertainties and volatilities to ensure higher unit value benefitting all the stakeholders across the value chain. The strategy of branding plantation commodities raises their value and does much to fetch a remunerative price. Brand is an intellectual property that distinguishes one product from another and it may be name, logo, trade mark, labelling, etc. The value of such legal devices is enormous and have commercial in nature. The commodity branding found to be very much relevant in both B2B and B2C markets, thus the brand owner may be individual farmer/planter, farmer's group or collectives, traders, processors, exporters and /or marketers.

Promotion of agriculture products on "country of origin labelling" (COOL) is one such branding initiatives shown much interest by the practitioners and widely studied by the academicians. One aspect related to COOL is traceability, a system allowing the product to be traced back to the producer is necessary in order to verify the source of origin of a product or production related credence characteristics. Labelling on food items act as signaling mechanism by which consumers can get assurances on the food quality standards, the place and the manner in which it is produced and processed. Information provided on food labels can influence buying behavior of the consumers. Several studies have demonstrated a link between the use of food labels and a dietary intake of the respondents. Food labels can be very valuable for consumers in order to make healthy and ethical food choices.

The impact of COOL on consumer behavior has been examined in the business and marketing literature for many years. Empirical studies show that COOL can affect consumers in a number of ways, including social status, store or product choice, perceived risk, and, in particular, product evaluation such as quality perception, product attitude or purchase intention. However, there is no such study carried out in India, measuring the effect of COOL on plantation commodities. The study has made an attempt to measure the Indian consumers' awareness level on coffee/ Tea, brand evaluation criteria, attitude towards foreign brands, ethnocentrism and their country's image and thereby to suggest a suitable branding initiative to promote these plantation products.

1.2. Indian Coffee Industry

India is the sixth largest producer of coffee in the world, accounting for over four percent of world coffee production. Indian coffee is grown in the states of Karnataka, Tamil Nadu and Kerala. There are over 1, 71,000 coffee farms in India, cultivating nearly 900,000 acres of coffee trees. Most coffee production in India is on small farms, with over 90 percent of all farms consisting up to 10 acres. The area under coffee cultivation is around 3,40,306 hectares, with a ratio of about 50/50 of Arabica and Robusta coffee. However due to the higher yields of this tree, Robusta accounted for 64 percent of all coffee produced in India. The annual production for the year 2013 - 14 is 304,500 MT. About 70 per cent of the produce is exported. Around 98 per cent of the plantations are owned by small growers who own less than 10 hectares of land.

The commercial plantations of coffee started in India during the 18th century. Coffee production in India grew rapidly in the 1950s, increasing from 18,893 tonnes in 1950-51 to 68,169 tonnes in 1960-61. Growth in India's coffee industry, however, has been especially robust in the post liberalisation era, backed by the government's decision to allow coffee planters to market their own produce, rather than selling to a central pool. Coffee production in India stood at 3,18,200 metric tonnes (MT) in 2012-13. Robusta variety accounted for 219,600 MT (69 per cent) of this production, while Arabica accounted for 98,600 MT (31 per cent). India has emerged as the seventh largest coffee producer globally; after Brazil, Vietnam, Columbia, Indonesia, Ethiopia and Honduras. It accounts for 2 per cent of the area under production and 3.7 per cent of the production in 2012 as compared to 3.18 per cent of production in 1992-93.

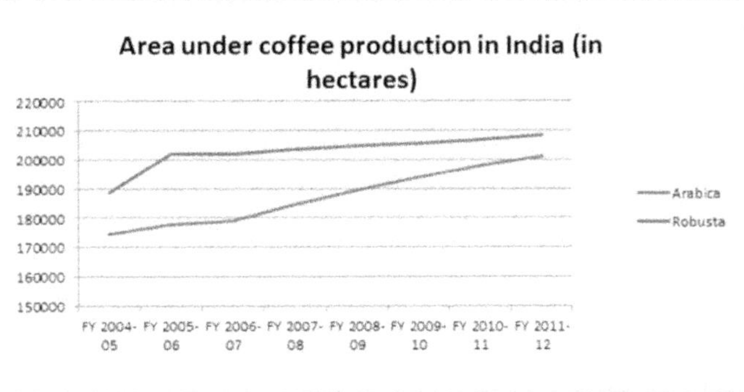

Fig. 1.1: Area under Coffee Production in India (in hectares)

The area under coffee plantations in India has increased by almost three times, from 120.1 thousand hectares in 1961 to 409.69 thousand hectares in 2012-13. Most of this area is concentrated in the southern states of Karnataka (72 per cent), Kerala (20.2 per cent) and Tamil Nadu (5.6 per cent). Productivity has also improved from around 567 kg/Ha in 1961 to around 852 kg/Ha during 2012-13. For the traditional areas, productivity has grown from 412 kg/Ha in 1961 to 944 kg/Ha in 2012-13. The industry is driven by the enterprise of around 2,80,241 coffee growers, out of which 99 per cent are small growers, while 1 per cent are medium to large growers. These plantations employ an average of around 6,06,702 people on an daily basis, as per estimates for 2011-12.

Coffee plantations in India use traditional methods and cultivate shade grown coffees under three-tiered canopies of wild and introduced trees. A lot of care is given to the selection of trees to be introduced. The primary shade or the lower shade is taken care of by nitrogen fixing Erythrina indica or by Glyrecedia maculata. These enrich the soil by harvesting atmospheric nitrogen and in turn give it to the coffee plant. The secondary shade is that of trees like silver oak, white and red cedar that shed their leaves in the monsoons and put forth a rich canopy during the summer. These trees are specifically selected because they act like biomass factories and thereby keep the soil temperatures low. Lastly, the tertiary shade is of the hardwood species, which attract rain bearing clouds. This three-tier shade system aids to filter harmful ultra violet radiation. Further, the filtered sunlight enables the sugars in the coffee bean to caramelize uniformly and give it a unique taste.

While growers are environmentally conscious at every step of cultivation, they are also beneficial, socially and economically to millions of farmers and allied households. Indian coffee is by and large associated with forest grown coffee and mechanization is kept to a bare minimum in Indian coffee plantations. Haphazardly arranged trees are characteristic of such plantations. The soil is virgin and efforts to mechanize the plantation are kept to a minimum in order to retain sustainable eco-friendly systems. The leaf litter from the trees acts like a sponge for the rainwater to absorb into the ground and this prevents runoff and soil erosion. This not only contributes to the soil fertility, but also protects the precious soil from weathering and other undesirable factors. Most coffee plantations are located in regions with average to heavy rainfall, yet even if there is a shower of 10 cms on one single day, there is no runoff inside the plantation because of the thick mulch, which acts as a blotting paper, allowing the water to slowly percolate downward. Mono-cropping is an exception in Indian coffee plantations and multicropping is popular. No other plantations in the world have the range of diversity that Indian coffee plantations have. This is owing to multiple, mixed cropping systems. Pepper vines are grown on shade trees, while cardamom, areca nut, ginger, citrus, vanilla and a few other spices are grown as multiple crops inside the coffee plantations. The matted roots of various crops that intertwine with one another culminate to give Indian coffee a spicy aroma. Indian coffees are often grown on terraced mountainsides. Indian arabicas (about half the crop) are known as plantation coffee, while the robustas are known as parchment coffee. Most coffee is grown in the Karnataka (Mysore) region, and Kerala and Tamil Nadu are the other main regions. Most of India's shade coffee comes from Karnataka, but the majority of India's arabicas are also shade grown. Another interesting coffee from India is 'monsooned' coffee made from green beans left exposed to monsoon rains in open warehouses. The beans turn tan colored, the acidity is reduced, and the beans are sweeter.

Most of the coffee plantations are situated in remote and uninhabited areas, where crops have not previously been grown. Hence, plantation workers are mostly migrants. These labourers are often provided housing in the estates itself along with facilities for shops, services and community activities such as recreation and cultural expressions. The Indian coffee industry witnessed a major crisis caused by falling coffee prices in 1990s. The fall in coffee prices led to closing down of production by small and medium sized coffee plantations, resulting in loss of jobs for thousands of coffee workers. The crisis also impacted wages paid to the workers, resulting in loss of income, food and clothing and medical facilities. Workers children stopped going to school and began to work along with them to augment family income. Many small growers are debt-ridden and few have even committed suicide due to falling prices and indebtedness.

Coffee is a globally traded commodity involving a large number of intermediaries in the value chain - farmer to consumer - unlike other plantation crops. Indian coffee created a niche in the international market and contributes significantly to the economy. Indian coffee is well received in European countries, USA, Japan, Russia, etc. The type of coffee produced in India with certain aroma and flavor components is highly sought after for blending purpose. Ironically, the Indian coffee loses its identity once gets blended. Moreover, Indian coffee does not enjoy international market acceptance like the Brazilian and /or Colombian coffee in spite of its unique quality characteristics.

The Indian coffee also faces stiff competition from Indonesia and Vietnam with their higher productivity and lower cost of production. Ultimately, the unit value realized per tonne of coffee is less.

Coffee Consumption in India

Before devising any strategy for growing domestic coffee consumption, the Coffee Board commissioned a comprehensive, nationally representative study on beverage habits, practices and attitudes. The third formal study was conducted by the Coffee Board in 2005. The report was commissioned to understand:

- Habits and practices with respect to coffee consumption of urban (South and North) and rural (South) consumers in India
- Coffee consumption by location and form
- Share of coffee in the daily basket of beverages consumed
- Attitudes towards coffee and drivers and barriers to coffee consumption
- Café habits

It covered all-India consumption by zones, age, gender and socio-economic classification (SEC) and attitudes towards coffee. The findings of the study revealed that the per capita consumption of coffee in India was 75 grams. The proportion of people consuming the beverage in the last 12 months increased in 2005 to 62 per cent from 59 per cent in 2003. 52 per cent preferred instant coffee and 15 per cent used R&G (filter coffee). About 94 per cent respondents consumed tea.

The study classified consumers into the following categories:

- Non-drinkers (38%)- did not consume coffee in the past 12 months
- Occasional drinkers (40%) - consumed coffee in the past 12 months but not yesterday
- Coffee drinkers (22%) - consumed coffee yesterday

On the basis of quantity consumed, the consumption was subdivided into light (1-2 cups a day); medium (3 cups a day) and heavy (4 or more cups a day).

It highlighted that the potential for growth of consumption lay with occasional drinkers and more so in North and Eastern zones where the occasional drinkers were maximum in number i.e. 64 per cent and 52 per cent respectively. Further, between 2003 and 2005, there was a marginal reduction of non-drinkers but the proportion of occasional drinkers remained the same.

The research also found that most of the daily consumption was at home. During 2005 about 24 per cent also consumed it away from home- a marginal increase of 2 per cent over the previous estimate of 22 per cent. Consumption of coffee away from home was mainly at restaurants (45%) and hot teashops (40%). Cafes and vending machines were also becoming more visible in the out-of-home segment. The study estimated that during the year 2005, the total volume of coffee consumption was 80,200 MT with an urban and rural divide of 58,500 MT (73%) and 21,700 MT (27%) respectively, representing an overall increase of 9900 MT or 7 per cent annual growth over the 2003 consumption of 70,300 MT (with an urban and rural share of 49,600 MT and 20,700 MT respectively). About 64,405 MT was consumed in south India, an 80 per cent share of all India consumption of 80,200 MT. In the south, Tamil Nadu was the largest coffee consuming state with an estimated volume of about 22,000 MT (34%) closely followed by Karnataka with 19,000 MT (30%). Andhra Pradesh and Kerala accounted for consumption of about 13,000 MT (20%) and 10,000 MT (16%) respectively. In the south zone, instant and filter coffee had an almost equal share. In other zones, instant coffee was more predominant.

Another study was commissioned by the Board on attitudes of consumers towards coffee with respect to quality, variety, price and additives especially chicory; positive and negative associations related to consumption as well as drivers and barriers to coffee consumption etc in 2006-07. It found that both coffee and tea enjoyed high spontaneous recall followed by plain milk and carbonated soft drinks. While 66 per cent of the respondents recalled tea, only 22 per cent recalled coffee as the first beverage. Top-of–the-mind recall of coffee was higher in traditional coffee strong holds such as Tamil Nadu and Karnataka. Across the country top-of-the-mind recall for coffee was higher among SEC A, B & C than SEC D (the lowest socio-economic strata). The study revealed that in Tamil Nadu, coffee had moderate association as a family beverage but was strongly associated with special occasions and hence had a higher status association. In Karnataka, coffee was very strongly associated as a family beverage. Interestingly, it also had special/high status perceptions. In Andhra Pradesh, perception of tea was very close to coffee on key dimensions. Tea came across mainly as an evening cup. In Kerala, tea had higher positive associations on key dimensions of family, health and status. In the North, East and West, coffee was largely a social drink. Coffee was associated with high status, modernity and antidote for cold weather suggesting it was a special and occasional drink. Health was a big barrier to more frequent consumption among those who drank only one cup of coffee a day. Among occasional coffee drinkers and non-drinkers, habit, non-consumption by other family members and price were key barriers. Taste was an additional barrier among non-drinkers in North, East and West and Tamil Nadu. Coffee as any other drink was habit-forming. Family drinking was a key for early adoption; most regular and occasional drinkers started drinking coffee at the age of less than 10 years at home, and were introduced to it by a family member. The exception was the North, where most were introduced to coffee by a friend and started consuming it outside home.

Growth of Domestic Coffee Market

India's per capita consumption of coffee was just 85 grams, compared to 4.5 kilograms in France, 4.6 kilograms in Japan and 6 kilograms in the US.[1] According to industry observer, India's coffee consumption pattern revealed that Indian market did hold enormous potential for growth. Coffee was preferred most only in the southern part of the country. Coffee consumption in India stagnated at 55,000 tons annually, till mid 1990s. The growing coffee café culture led by domestic brewers such as CCD and Barista resulted in enormous growth in coffee consumption. In fact, it had doubled since 1990s. Over the years, other coffee chains such as the Coffee Bean & Tea Leaf, The Chocolate Room, Qwiky's and Café Nescafe entered the Indian market.

Hindustan Unilever Ltd., the Indian subsidiary of consumer packaged goods behemoth Unilever, was pilot-testing Bru World Café in Mumbai.

The domestic consumption for coffee is growing at a rate of 5-6 per cent annually since 2010. "The sector, with the Coffee Board of India, is actively working on many initiatives to spread awareness about coffee. One is the India International Coffee Festival (IICF). It plays an important role in promoting consumption in India. There are signs that the popularity of coffee is increasing with the spread of both foreign and home-grown coffee shops and restaurant chains. Western-style

1 Suparna Goswamy Bhattacharya, —Starbucks Could Cause A Veritable Tempest In Local Coffee Mart,‖http://www.dnaindia.com/money/report, April 11, 2011

coffee bars have sprouted all over India, as a newly emergent middle class has become enamored with café latte and other such caffeinated delicacies. Chains like Coffee Cafe Day, Barista and others have a widespread presence in all Indian cities, which makes it evident that people in India like coffee. Coffee culture "has taken India by storm and it will continue to do so because of higher incomes and greater urbanization."

To cash in on the growing consumption of coffee in the domestic market, the companies like Tata Coffee is looking at vending solutions to offer innovative products and services. The Indian market became home to multiple coffee brands including Café Coffee Day, Barista Lavazza, Java Green, Costa Coffee, and Gloria Jean's Coffee, but the market was still in a nascent stage. India was the sixth largest coffee producer in the world with 4 percent share of global coffee production. But, it consumed relatively a very tiny volume for a population of 1.2 billion. Indian market was estimated to grow by at least 100 million new coffee drinkers in the near future. These predictions would result in potential for more international and domestic coffee chains' entry into the country.

1.3. Indian Tea Industry

Tea had been known for millennia in India as a medicinal plant, but was not drunk for pleasure until the British began to establish plantations in the 19th century. Darjeeling tea, from the Darjeeling region in West Bengal has traditionally been prized above all other black teas, especially in the United Kingdom and the countries comprising the former British Empire. The Chinese variety is used for Darjeeling tea, and the Assamese variety, native to the Indian state of Assam, everywhere else. The British started commercial tea plantations in India and in Ceylon. In 1824 tea plants were discovered in the hills along the frontier between Burma (Myanmar) and the Indian state of Assam. Currently grown in 35 countries, the tea industry provides a source of employment and export earnings for the developing and underdeveloped nations.

The British introduced the tea culture into India in 1836 and into Ceylon (Sri Lanka) in 1867. India was the top producer of tea for nearly a century, but was displaced by China in the 21st century. Indian tea companies have acquired a number of iconic foreign tea enterprises including British brands Tetley and Typhoo. Tea is cultivated in the high ranges of North and South India and the best quality is known as CTC and Orthodox Assam Tea, respectively. The consumption is above the 600 million kg mark per year. The market consists of both leaf and dust teas both in the CTC and Orthodox Grades, with the southern markets consuming more dust teas. Nearly every part of the country has a tea-growing region. Approximately 4 per cent of the national income of India comes from its tea and India is home to over 14,000 tea estates. The geography of India allows for many different climatic conditions, and the resulting teas can be dramatically different from each other. There are three main kinds of tea produced in India:

Assam tea: Assam tea comes from the north-eastern region of the country. This heavily forested region is home to much wildlife, including the rhinoceros. Tea from here is rich and full-bodied. It was in Assam that the first tea estate was established, in 1837.

Darjeeling tea: The Darjeeling region is cool and wet, and tucked in the foothills of the Himalayas. The tea is exquisite and delicately flavored, and considered to be one of the finest teas in the world. The Darjeeling plantations have three distinct harvests, and the tea produced from each 'flush' has a unique flavor. First flush teas are light and aromatic, while the second flush produces tea with a bit more bite. The third or autumn flush gives a tea that is lesser in quality.

Nilgiri tea: This tea comes from an even higher part of India than Darjeeling. This southern Indian region has elevations between 1,000 and 25,000 meters. The flavors of Nilgiri teas are subtle and gentle. They are frequently blended with other, more robust teas.

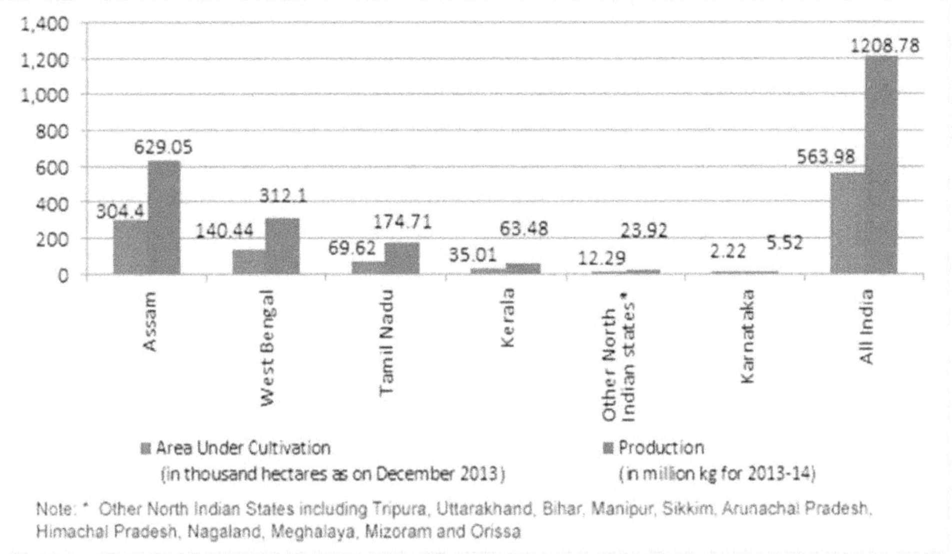

Fig. 1.2: Area under Tea Cultivation and the Production in India

By building on a proud legacy of enterprise that spanned nearly two and a half centuries, India has acquired an exalted status on the global tea map. The country is the second largest tea producer in the world with production of 1205.40 million kg in 2013-14. Interestingly, India is also the world's largest consumer of tea with the domestic market consuming 890 million kg of tea during 2012-13. India is ranked fourth in terms of tea exports, which reached 218.12 million kg during 2013-14 and were valued at US$ 727.04 million. The top importing markets were Commonwealth of Independent States (CIS, 51.58 million kg), UAE (23.19 million kg) and Iran (22.42 million kg). All varieties of tea are produced by India. While

CTC accounts for around 89 per cent of the production, orthodox/green and instant tea account for the remaining 11 per cent.

India has around 579.35 thousand hectares of area under tea production. Tea production is led by Assam (322.21 thousand hectares), West Bengal (115.10 thousand hectares), Tamil Nadu (80.46 thousand hectares) and Kerala (37.14 thousand hectares). According to estimates, the tea industry is India's second largest employer. It employs over 3.5 million people across some 1,686 estates and 157,504 small holdings; most of them women.

With an export of approximately 210 million kg of tea, India stands as the fourth-largest exporter of tea in the world, with China at the top. About 72 per cent of the country's tea sector is organized. While India is the largest consumer of tea worldwide, the per-capita consumption of tea in India remains a modest 750 grams per person. The global consumption of tea jumped 60 per cent between 1993 and 2010, and significant growth is forecast, as more people become consumers of tea. The total tea production in the world has exceeded four billion kg, with India producing about one billion kg of tea. Between 2008 and 2013, black tea production in India increased at a compounded annual growth rate (CAGR) of 1.6 per cent, while consumption rose at a CAGR of 2.3 per cent.

Tea Consumption in India

India is one of the largest tea producers in the world, with over 70 per cent of the tea being domestically consumed. The Indian tea industry has grown to own many global tea brands, and has evolved to one of the most skilled tea industries in the world. The market size of tea is estimated to be approximately Rs 10,000 crore, with a penetration of more than 90 per cent in the domestic market.

Per capita consumption rose to 718g (1.58lb) in 2011, which was about triple the domestic consumption in the United States. It rose steadily from a per capita count of 654g (1.44lb) in 2001. We can see the diversity in the consumption pattern in India, the way in which tea is consumed and embraced by both the rich and influential and commoners, with variations depending on regional and cultural affiliations.

1.4: Usage of labels for as Brand Promotion and Consumer Protection

Food product labelling, as policy tool for ensuring provision of nutrition and health information to consumers and as product differentiation strategy by food companies, has gained importance in the recent past across the globe (Kim, Nayga and Capps, 2001; Marks, 1984)[2]. Recent concerns with food safety have resulted in increased demand for regulatory pressures directed at labelling the ingredients, processing methods, and nutritional content of foods. Trend towards

2 Kim S.Y., Nayga R.M. Jr., and Capps O. Jr., 2001. Food Label Use, Self-Selectivity, and Diet Quality. Journal of Consumer Affairs, 35(2): 346-348

healthier and wellness food has also led to consumer demand for "more detailed, accurate, and accessible" nutritional information on the packaged food (Abbott, 1997)[3]. The food labels act as a signaling mechanism by which food companies assure their potential consumers regarding their sound quality control practices. Having a supportive marketing environment that provides content of food items can be considered as a principle in promoting the health of consumers. Providing food content information on packets can be thought of as an important element for consumer protection. "Consumers have as much right to know the nutrient content of the foods they choose to purchase as they do to know its country of origin and that it is safe to eat" (Cowburn and Stockley, 2005)[4]. "Labelling includes any written, printed or graphic matter that is present on the label, accompanies the food, or is displayed near the food, including that for the purpose of promoting its sale or disposal". Many countries in the world are developing mandatory or voluntary programs to assure food safety by using traceability in food value chain "Traceability is defined as the ability to follow the movement of a food through specified stage(s) of processing, production, and distribution" (Souza-Monteiro and Caswell, 2004)[5]. Therefore, to come out with a suitable strategy to label coffee/ tea with an aim of educating and protecting the Indian consumer through brand promotion, the contributing antecedents need to be studied in detail.

1.5: Consumer Ethnocentrism, Country Image and Country of Origin Effect: A Conceptual Framework

Consumer Ethnocentrism and Purchase Intention

Ethnocentrism was originally a sociological concept. Sumner defined ethnocentrism as 'view of things in which one's own group is the center of everything, and all others are scaled and rated with reference to it.

According to Levine and Campbell, ethnocentrism was originally a sociological concept, which then became a psychosocial construct with relevance to individual-level personality systems, as well as to the more general cultural- and social-analytic frameworks.

Shimp and Sharma expanded this concept to consumer behavior and created the CETSCALE (Consumer Ethnocentrism Tendencies Scale). Consumer ethnocentrism is defined as the beliefs (knowledge structures and thought processes) held by consumers about the appropriateness, indeed morality, of purchasing foreign-made products in place of domestic ones.

---

3 Abbott R., 1997 Food and Nutrition Information: A Study of Sources, Uses, and Understanding. British Food Journal, 99(2): 42-44

4 Cowburn G, Stockley L (2005). Consumer understanding and use of nutrition labelling: a systematic review. Public Health Nutr. 8(01):21-28.

5 Souza-Monteiro DM, Caswell JA (2004). The economics of implementing traceability in beef supply chains: Trends in major producing and trading countries University of Massachusetts, Amherst Working Paper No. 2004-06.

The best explanation of the concept is in the words of the originators of the concept (Shimp and Sharma, 1987): We use the term 'consumer ethnocentrism' to represent the beliefs held by American consumers about the appropriateness, indeed morality, of purchasing foreign-made products. From the perspective of ethnocentric consumers, purchasing imported products is wrong because, in their minds, it hurts the domestic economy, causes loss of jobs, and is plainly unpatriotic; products from other countries (i.e., out groups) are objects of contempt to highly ethnocentric consumers. To non-ethnocentric consumers, however, foreign products are objects to be evaluated on their own merits without consideration for where they are made (or perhaps to be evaluated more favorably because they are manufactured outside the United States).

Previous research on the effects of consumer ethnocentrism on consumer behavior has revealed that high-ethnocentric consumers take unreasonably favorable evaluations of domestic products vis-à-vis imported products (Bilkey and Nes 1982; Han and Terpstra 1988; Johansson *et al.* 1985; Sharma *et al.* 1995; Wall and Hcslop 1986; White 1979).

The consumers with higher consumer ethnocentrism have better attitude toward domestic products. When consumers have positive product attitude, it means that they judge the products better and have stronger purchase intention. On the other hand, the subjective norm is a personal internal trait which reflect one's nature, while consumer ethnocentrism only represent one's attitude toward purchasing foreign products. Hence consumer ethnocentrism has no direct impact on subjective norm. Based on the above discussion, Fig. 1.3 shows the relationship among four variables. Attitude and subjective norm have direct influence on purchase intention, while consumer ethnocentrism has indirect impact on purchase intention through product attitude.

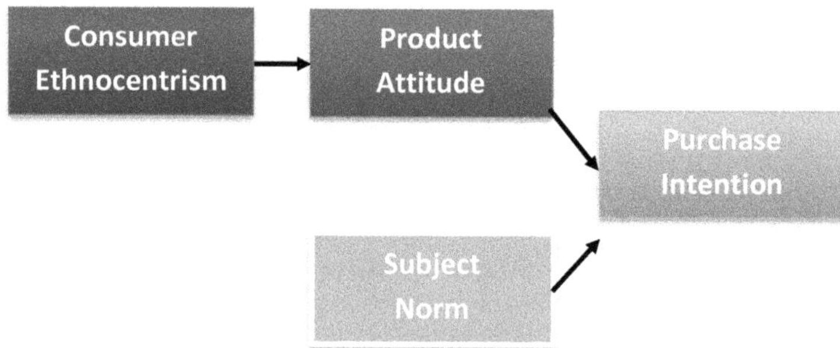

Fig. 1.3: Relationship between Consumer Ethnocentrism & Purchase Intention

Consumer ethnocentrism has significant positive impact on product evaluation and attitude. Product attitude have positive impact on purchase intention.

With the growth of international trade and travel, consumers are increasingly confronted with foreign products and services. But some negative attitudes towards foreign products can arise from several factors such as previous or ongoing political, military, economic, or diplomatic events. Thus, both consumer ethnocentrism and consumer animosity have become important constructs in marketing.

Country Image and Consumers' Perception

Image, as a simple explanation, is an idea and the value judgments of a target audience on a specific country, nation, subject or a product. It is how the audience shape the publicity in their mind. Continuing a positive image is important to all companies. If the image is positive, consumers expound everything they live with the companies positive, but if the image of the company is negative; even if they aren't true the company as a whole can badly be affected. Wrong perceptions have a great damage on the image of the companies. Positive and negative corporate image, has great influence on company's profit and loss accounts, while in macro programing, a country's or a nation's images has impacts on encouraging trade, tourism and inward investments of that country. In that perspective, a positive national image becomes an essential for all countries. Country image was first defined by Nagashima as "the picture, the reputation, and the stereotype that businessmen and consumers attach to products of a specific country. This image is created by such variables as representative products, national characteristics, economic and political background, history, and traditions. Country image is the overall perception consumer's form of products from a particular country, based on their prior perceptions of country's production and marketing strengths and weaknesses. Despite a large body of literature on the subject COO, the number of studies that have in fact included country image measures remains very limited.

One of the first studies explicitly focusing on country image measures was conducted by Martin and Eroglu (1993, p. 193), who defined country image as "the total of all descriptive, inferential and informational beliefs one has about a particular country." According to Martin and Eroglu (1993), COI is a three-dimensional construct consisting of a political, an economic as well as a technological dimension. Whereas these dimensions clearly reflect consumers' cognitive perceptions about a country, past research has shown that "country of origin is not merely a cognitive cue for product quality, but also relates to emotions, identity, pride and autobiographical memories" (Verlegh & Steenkamp, 1999, p. 523). A number of authors (e.g. Obermiller & Spangenberg, 1989; Papadopoulos, Heslop, & Beracs, 1989; Parameswaran & Pisharodi, 1994; Verlegh & Steenkamp, 1999; Laroche, Papadopoulos, Heslop, & Mourali, 2005) therefore suggest that the construct of country image is comprised of:

- a cognitive component, which includes consumers' beliefs about the country's industrial, technological as well as political background;
- an affective component that describes the country's symbolic and emotional value to the consumer, and
- a conative component, capturing consumers' desired interaction with the sourcing country.

While it seems to be commonly accepted that images should consist of these three dimensions, the majority of the existing COI studies does not include all three facets. Furthermore, most of these studies have used two items only, i.e. "people are friendly and likeable" or "people are trustworthy" to measure the affective component of country image. Thus, cognitive measures prevail. Furthermore, these items could represent emotions as well as cognitions, which make it difficult to classify them into one of these facets.

Country of Origin Effect and Product Evaluation

Consumer perceptions on the country of origin effect play a major role in influencing a consumer's choice of a product. Research in international marketing has proven that country associations do lead to customer bias. Country of origin effect can be defined as any influence that the country of manufacture has on a consumer's positive or negative perception of a product (Cateora and Graham, 1999). With increasing availability of foreign goods in most national markets, the country of origin cue has become more important as consumers often evaluate goods differently than they do competing domestic products. (Bilkey and Nes, 1982). Over the past three decades, the effect of a product's country of origin on buyer perception and evaluation has been one of the most widely studied phenomena in international business, marketing and consumer behavior. With the increasing pace of globalization and the diversity of manufacturing activities internationally, more studies are needed to guide marketers to have a better insight into buyers' attitudes and behavior with respect to global products.

Country-of-origin image (COI) is an important driver of consumers' evaluation of products originating from different countries. Schooler (1965) is generally considered to be the first researcher to empirically study this effect. He found out that products, identical in every respect except for their country of origin, were evaluated differently by consumers. Since then, more than 700 articles have been published on the subject country-of-origin (Papadopoulos & Heslop, 2003).

Consumers use both intrinsic and extrinsic informational product cues as the basis for their evaluation of products (Ulgado & Lee, 1998). Intrinsic cues involve the physical composition of a product, whereas extrinsic cues are product related, but are not part of the physical product itself. Brand name, retailer reputation, and products' country of origin are regarded as extrinsic cues and can be manipulated without physically changing the products (Verlegh & Steenkamp, 1999).

Realizing that consumers may use one of the extrinsic cues: i.e., country of origin, as a signal to infer beliefs regarding product attributes such as quality, researchers mainly studied the use of country of origin as a cognitive cue (Steenkamp, 1990). The predictive value of such a cue is affected by either the "ecological" or "observed" covariation between cue and attribute (Steenkamp, 1989), or by the theoretical or intuitive relations hip between cue and attribute (Pinson, 1986).

Research has shown that country of origin serves as a signal for product quality and performance. Erickson, Johansson and Chao (1984) developed a model that involves country of origin and other product attributes such as quality and performance. They found a "halo effect" of country of origin: that is, country image affects beliefs about tangible product attributes, and in turn affects overall evaluation.

When making buying decisions, consumers may link country of origin to personal memories, to national identities and to feelings of "pride" associated with the possession of products from certain countries (Hirschman, 1985).

Realizing that consumers may use one of the extrinsic cues: i.e., country of origin, as a signal to infer beliefs regarding product attributes such as quality, researchers mainly studied the use of country of origin as a cognitive cue (Steenkamp, 1990). The predictive value of such a cue is affected by either the "ecological" or "observed" covariation between cue and attribute (Steenkamp, 1989), or by the theoretical or intuitive relations hip between cue and attribute (Pinson, 1986). The perceived theoretical relationship between the cue of country of origin and the attributes of a product is largely conducted by product-country images, among which quality as a representation of a country's production has an important effect on consumers' evaluations of products

(Broniarczyk & Alba, 1994). A preference for German cars, for example, may be explained by the perception of advanced technological quality of the German industry as a whole.

Research has shown that country of origin serves as a signal for product quality and performance. Erickson, Johansson and Chao (1984) developed a model that involves country of origin and other product attributes such as quality and performance. They found a "halo effect" of country of origin: that is, country image affects beliefs about tangible product attributes, and in turn affects overall evaluation. Also, Han (1989) found that when unfamiliar with a country's product, consumers infer product information into country image, which then influences consumers' attitudes toward other attributes.

Branding and Promoting Indian Tea and Coffee

The studies on consumer ethnocentrism reveals that as the country develops, literacy rate increases, and people travel more, etc., the CET score comes down. Therefore, Indian tea and coffee having unique characteristics and demanded globally can very much be branded on country of origin label instead of a commodity. Branding is not just creating fancy names and logos. It is a strategic tool for differentiating an organization in the market place and building up competitive advantage. The branding initiative will help in strengthening the competitive position of Indian tea and coffee in the global market in several ways. It will add value to the product beyond its functional purpose, thereby it supports the price, it serves as a link between the Indian tea and coffee planters with the customers, enhancing the farmers' share in the price spread and thereby facilitates in enhancing the market efficiency.

The country-related and consumer-related factors that drive behavioral intentions thus need to be considered when promoting products on COOL. The model developed by Katharina Petra Roth (2006) for the promotion of tea and coffee using COOL is diagrammatically shown in the following figure.

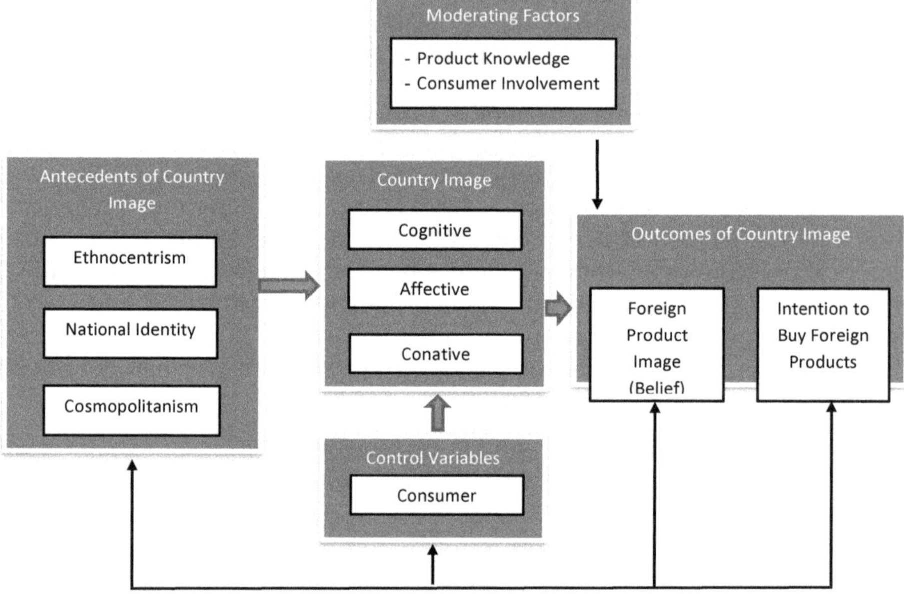

Fig. 1.4: Framework for Promotion on Country of Origin (adapted from Katharina Petra Roth: 2006)

The stakeholders of Indian tea and coffee such as planters, traders, processors, exporters and policy makers, intending for promoting the produce internationally using country of origin labelling for market competitiveness, therefore, need to understand the consumers' level of awareness on product quality, where and how coffee/tea is produced, consumers' ethnocentrism and the perceived country image.

2

CHAPTER

Review of Literature

2.1. Consumer Ethnocentrism

The term "consumer ethnocentrism" is adapted from the general concept of ethnocentrism introduced more than 80 years ago by Sumner (1906). From the perspective of ethnocentric consumers, purchasing imported products is wrong because, in their minds, it hurts the domestic economy, causes loss of jobs, and is plainly unpatriotic; products from other countries (i.e., out groups) are objects of contempt to highly ethnocentric consumers. To non-ethnocentric consumers, however, foreign products are objects to be evaluated on their own merits without consideration for where they are made. There are several studies pertaining to the consumer ethnocentrism and brand evaluation and preference formation. Summary of few relevant studies are compiled below.

Kaynak and Kara (2002): [6]According to them, highly ethnocentric individuals are intolerant of cultural differences of other nations and they perceive ethnic and national symbolic values as sources of pride while often despising the values of others. They found that consumers scoring high on Consumer Ethnocentric Tendency (CET) were less willing to buy foreign products and that the degree of CET varied considerably across consumers and regions of the country. They also show that CET is correlated to various lifestyle dimensions and that it differentially affects foreign local product dispositions, depending on the nature of the product category and the products country of origin.

Shimp & Sharma (1987):[7] According to them, highly ethnocentric customer's negatively evaluated imported products due to its negative impact on the domestic economy and positively correlated towards domestic products. The consumer

6 Kaynak E & Kara, A. (2002). Consumer's perceptions of foreign products. European Journal of Marketing, 36(7/8), 928-949. http://dx.doi.org/10.1108/03090560210430881

7 Shimp, T & Sharma, S.(1987). Consumer ethnocentrism: construction and validation of the cetscale. Journal of Marketing Research, 14, 280-289. Retrieved from http://jstor.org

ethnocentric tendency was measured using the famous CETSCALE. Furthermore senior citizens are most ethnocentric than the youth because of their conservative nature. More ethnocentric consumer tends to prefer home made products, based on morality they attach to the purchase of foreign made ones. Highly ethnocentric consumers perceive that purchasing foreign-made products is wrong due to the damage it causes to the local economy, i.e. creating unemployment, which plainly labels them as unpatriotic.

Shankarmahesh (2006):[8] According to him, the main categories of consequences are attitudes towards foreign products, purchasing intention and support for foreign products. Additionally he suggested that perceived equity, empathy, perceived cost, responsibility, country of Origin and product evaluation as mediating variables between CE and its outcomes. In his thesis" Impact of demographic variables on consumer ethnocentric tendencies, suggested the Country of origin and product evaluation as mediating variables between CE and its outcomes.

Lee, Hong & Lee (2003):[9] According to them, CE is used by customers to predict judgments of the quality of foreign products. Generally highly ethnocentric customers are motivated to buy domestic products even if they know the quality of the products is lower than the imported alternatives.

Juric *et al.*, **(1995):** [10]His study revealed that higher the ethnocentric tendency the more negative the attitudes towards foreign or imported brands will be and thus more positive they are towards domestic products. Accordingly, he hypothesized that the consumer ethnocentric tendency is negatively associated with favorable attitudes towards foreign brands and Consumers exhibiting high levels of ethnocentrism will have less favorable attitudes towards foreign brands.

Shimp, Sharma and Shin (1995): [11]Their study revealed that the global forces of marketing and media indeed convey consumption values and behaviors and it could be constructed that highly ethnocentric consumers would eschew such global influences and therefore be less concerned about material possessions. **De Mooij (2004):**[12] According to him, the cross cultural differences in product ownership

8 Shankarmahesh M.N (2006). Consumer ethnocentrism: an integrative review of its antecedents and consequences. International Marketing Review, 23(2), 146-172. http://dx.doi.org/10.1108/02651330610660065.

9 Lee, W., Hong,& Lee, S (2003): Communicating with American consumers in the post 9/11 climate. An empirical investigation of consumer ethnocentrism in the United States. International Journal of Advertising, 22, 487-510. Retrieved from http://ehis.ebscohost.com/

10 Juric *et al.* (1995): : Open market aftershocks: Czech and Slovak attitude towards brands". Towards A market economy: Beyond the Point of No return, Second east and central European conference, ESOMAR, Warsaw, April 1995.

11 Shimp, Sharma Shin (1995): Consumer Ethnocentrism: A test of Antecedents and moderators," Journal of the Academy of Marketing sciences,23 (1), 26-37.

12 De Mooji (2004): Consumer Behavior and Culture: Consequences for Global Marketing and Advertising. Thousand Oaks, CA: Sage Publications.

and usage could largely be attributed to the link between product category and cultural values. The older the product category, the stronger is the influence of traditional culture. Consumer ethnocentrism should be most prominent for product categories that are highly culture bond.

Vida and Fairhust (1999):[13] Their dominant view is that older will be more consumers ethnocentric than the younger and regarding education, there is a near consensus that higher the education level lesser the consumer ethnocentrism. **Agrawal (1994):** [14] According to him, the desire for foreign products could be quest for quality especially as preference for foreign goods is higher in the less developed countries and more among higher income categories. The different socio demographic groups will differ with respect to consumer ethnocentrism. The most consumer ethnocentric group will have higher average age, least education, least average income, significantly more members from the lower SEC grade, significantly more females and will be least quality conscious.

Summer (1996):[15] According to him, ethnocentrism is "the technical name for the view of things in which one's own group is the center of everything, and all others are scaled and rated with reference to it". In the other words, ethnocentrism is a tendency of people to believe that their cultural or ethnic group is important and the other groups are evaluated primarily from the perspective of one's own culture. **Lutz** *et al.* **(2008):**[16] According to him, ethnocentric consumers do not intend to buy foreign-made goods, since they think it is harmful to the domestic economy. Foreign presence may have negative competition effects on domestic firms. It would disadvantage domestic producers, increase unemployment, and worsen economic conditions in the home country. They suggested that "the consequences of consumer ethnocentricity include overestimation of the quality and value of domestic products and underestimation of the virtues of imports, a moral obligation to buy domestic products, and intense preference for domestic products".

Kucukemiroglu (1999): [17]In his hypothesis, found that non-ethnocentric consumers tend to have significantly more favorable beliefs, attitudes and intentions regarding imported products than ethnocentric consumers. Thus, based on this existing

---

13 Vida and Fairhust (1999): Factors underlying the phenomenon of Consumer ethnocentricity. Evidence for four central European countries. "The international review of Retail distribution and consumer research, 90(4),321-337

14 Agarwal(1994): Marketing Research, Delhi. Global Business Press

15 Sumner, W. G. (1996). *Folkways: The sociological importance of usages, manners, customs, mores, and morals.* NY: Harper & Row.

16 Lutz,S, Talavera, O & Park, S.M (2008). Effects of foreign presence in a transition economy, regional and industrywide investments and firm level exports in Ukaranian manufacturing, emerging markets finance and trade,44 (5),82-98

17 Kucekemiroglu,o (1999):Market segmentation by using consumer lifestyle dimensions and ethnocentrism. An empirical study, European Journal of marketing, 335(6),470-487.

body of previous research, the following hypothesis is proposed, "Ethnocentric attitudes of consumers have a positive effect on Domestic Purchase Behavior and a negative effect on Foreign Purchase Behavior".

Attitude Towards Foreign Products

The influence of foreign brands on consumer behavior is a hot topic. It is first studied in the literature in terms of brand strategy to be adopted by enterprises in the international market. **Motameni & Shahrokhi (1998)**[18] have noted three main points to understand: 1. How consumers in each country consider the choice of brands, 2. How they evaluate them, 3. The different needs that drive their purchasing decision. In a context of globalization of markets, there are many studies about the alternatives of a marketing strategy, standardized or localized, which derives the question of a branding strategy standardized globally vs brand adapted (Kapferer, 2002, Roth, 1992, and 1995). Research shows that it is important to identify potential obstacles, for example, some national characteristics, in the process of developing a global brand.

In examining the literature on consumer attitudes towards foreign products, two principal streams of research can be identified at first. One focuses on the impact of country of origin (COO) on consumer attitudes and specifically, its use as a cue in making inferences about or evaluating foreign products. This often generated conflicting and ambiguous findings. The second strand focuses on factors underlying attitudes towards foreign products, such as the impact of hostile attitudes towards a specific country, and the effect on buying intent and ownership.

A large body of literature proposes that consumers in emerging markets prefer foreign brands to local products to obtain social status, social conformity, and wealth expression (Batra *et al.* 2000; Ger and Belk 1996; Wang and Yang 2008; Zhou and Hui 2003). **Batra** ***et al.*** **(2000)**[19] argue that symbolic motives, such as social status display and communication, primarily influence the generalized preference for foreign brands in developing countries characterized by low-level economic development and a high degree of social mobility

Lantz & Loeb (1996):[20] According to them, most of the brands characteristics are associated with its country of Origin. They explored that the COO is more important than the brand name, price and quality in shaping attitudes towards

18 Motameni R. & Shahrokhi M. (1998), Brand equity valuation: a global perspective, Journal of Product and Brand Management, 7, 4, 275-290

19 Batra, Rajeev, Venkatram Ramaswamy, Dana L. Alden, JanBenedict E.M. Steenkamp, and S. Ramachander (2000), "Effects of Brand Local and Nonlocal Origin on Consumer Attitudes in Developing Countries," Journal of Consumer Psychology, 9 (2), 83–95.

20 Lantz & Loeb (1996): Country of Origin and ethnocentrism: an analysis of Canadian and American preferences using social identity Theory. Advances in Consumer Research, 23, 374-378. Retrieved from http://ehis.ebscohost.com/

the particular Products. **Kaynak & Cavusgil (1983):**[21] They argued that domestic customers may develop favourable attitudes towards the products that come from countries with similar cultural, political and economic conditions.

Adorno *et al.* **(1950):**[22] According to him, scores obtained on the CET scale were correlated with five other scores- belief about foreign made products, belief about Indian made products, attitude towards foreign products, attitude towards Indian products and the importance of buying products made in India. They found that the consumer ethnocentric scores correlated negatively with attitude towards foreign products, they correlate positively with attitude towards Indian products. It was expected that the correlations of CET SCORE will be higher with attitudes (both for foreign and Indian made products) than with beliefs (both with Foreign and Indian made products.

Huddleston, Linda & Lesli (2001):[23] According to them, the COO of imported brands should be considered as a primary factor in assessing the attitudes towards foreign brands. They have attempted to develop measurement standards to identify to what extent customers value extrinsic and intrinsic cues of domestic brands compared to foreign alternatives. **Shimp & Sharma (1987):**[24] According to them, highly ethnocentric customers negatively evaluated imported products due to its negative impact on the domestic economy. They further argued that not so strong ethnocentric customers show a favorable attitude towards imported products just because of its foreign origin. Furthermore, highly ethnocentric customers do not make purchasing decisions based on perceived brand globalness compared to not so strong ethnocentric customers.

Mooij & Hofstede (2002):[25] They suggested that consumer behavior in global marketplace will not be homogeneous due to cultural differences. The transitional economies of the CIS countries have not been researched while these countries raise an interesting question as to the differential effects of attitudes and purchase behavior between the once dominant regimes versus the controlled regime.

---

21 Kaynak & Cavusgil (1983): Consumer attitudes towards products of foreign origin: do they vary across product classes? International Journal of advertising,2,147-157.

22 Adorno *et al.* 91950): The authoritarian Report.

23 Huddleston, P. Linda, K & Lesli, S (2001): Consumer ethnocentrism, product and polish consumers perceptions of quality. International Journal of retail and distribution management, 29(5), 236-246. http://dx.doi.org/10.1108/09590550110390896.

24 Shimp & Sharma (1987): Consumer ethnocentrism: Construction and validation of the cetscale. Journal of marketing research,14,280-289. Retrieved from http://jstor.org

25 Mooki & Hofstede (2002): The roles of consumer ethnocentricity and attitude toward a foreign culture in processing foreign and origin advertisements. Advances in consumer research, 23,436-439.

Lee, Hong & Lee (2003):[26] According to them, CE is used by customers as predict judgments of the quality of foreign products. Generally, highly ethnocentric customers are motivated to buy domestic products even if they know the quality of the products is lower than the imported alternatives. Moreover, those consumers perceived that foreign products are of a lower quality. Furthermore, they noted that highly ethnocentric customers are not motivated to learn about the attributes of foreign brands. Most of their studies have revealed that higher the ethnocentric tendency, the more negative the attitudes towards foreign or imported brands will be, and thus more positive they are towards domestic products. They suggested that Consumer ethnocentric tendency is negatively associated with favorable attitudes towards foreign brands and high levels of ethnocentrism will have less favorable attitudes towards foreign brands.

Good & Huddleston (1995):[27] They noted that older customers have negative attitudes towards foreign products. Furthermore argued that price, taste and healthiness of products are more important irrespective of age, gender and education in developing attitudes towards foreign brands. They further emphasize that educated customers maintain favorable attitudes towards foreign brands. The income level of customers positively correlates with the favorable attitude towards foreign products. Moreover, males have more positive attitudes towards domestic brands than female customers. They found that there is a significant relationship between age, gender, income and education (demographic characteristics) with their degree of ethnocentrism.

2.2. Country Image

One of the first studies explicitly focussing on *country* image measures was conducted by Martin and Eroglu (1993, p. 193), who defined country image as *"the total of all descriptive, inferential and informational beliefs one has about a particular country."*

Martin and Eroglu (1993):[28] According to them, Country of Origin (COI) is a three-dimensional construct consisting of a political, an economic as well as a technological dimension. Whereas these dimensions clearly reflect consumers' cognitive perceptions about a country, past research has shown that *"country of origin is not merely a cognitive cue for product quality, but also relates to emotions, identity, pride and autobiographical memories"*. **Kaynak, E., Kucukemiroglu, O., &**

26 Lee, W.Hong, J & Lee, S (2003): Communicating with American consumers in the past 9/11 climate: An empirical investigation of consumer ethnocentrism in the United states. International Journal of advertising, 22,487-510. Retrieved from http://ehis.ebscohost.com/

27 Good, L. K., & Huddleston, P. (1995). Ethnocentrism of polish and Russian consumers: Are feelings and intentions related? International marketing review, 12(5), 35-48. http://dx.doi. org/10.1108/02651339510103047.

28 Martin, I.M & Eroglu, S (1993): Measuring a multi-dimensional construct: Country image. Journal of business research,28(3),191-210.

Hyder, A. (2000):[29] According to them, the country of origin effect (COO) and relative product quality perceptions of domestic goods versus foreign products could be important determinants of consumer behavior.The other limitation of this study is that price of domestic vs. imported goods was not controlled. It is hard to determine whether consumers purchase domestic goods because they have high ethnocentric attitudes or because the domestic goods are simply cheaper than imported ones. The managerial implication is that foreign firms should assure domestic consumers that purchasing of their products would not disadvantage domestic producers and worsen economic conditions in the home country.

Klein *et al.* **(1998) and Balabanis and Diamantopoulos (2004):**[30] They suggested that the CE constructs appear to be more capable of explaining consumers' positive bias toward home products rather than negative bias against foreign products, pointing out the weaknesses of ethnocentrism in providing specific insights regarding consumer aversion toward a foreign country. **Kayank** *et al.* **(2000):** He hypothesized that, the country of origin effect (COO) and relative product quality perceptions of domestic goods versus foreign products could be important determinants of consumer behavior. It is hard to determine whether consumers purchase domestic goods because they have high ethnocentric attitudes or because the domestic goods are simply cheaper than imported ones. Finally, a more sophisticated sampling procedure can help to study the relationship between ethnocentrism and consumer characteristics.

Chen and Chang (2008):[31] According to them, by incorporating brand preference and purchase intention as consequences of brand equity a conceptual model was developed where brand equity is treated as the independent variable, brand preference and purchase intention treated as the dependent variables, also COO considered as moderated variable.

Chang and Liu (2009):[32] Based on the conceptual framework model 5 hypothesis was examined by them.

H1. Consumer-based brand equity has direct and positive influence on consumer's brand preference.

H2. Consumer-based brand equity has direct and positive influence on consumer's purchase intention.

---

29 Kaynak, E.Kucukemiroglu, O & Hyder, A (2000).Consumers Country of origin COO perceptions of imported productsin a homogenours less developed country. European Journal of marketing, 34 99),1221-1241.

30 Klein *et al.* (1998): Market segmentation by using consumer lifestyle dimensions and ethnocentrism. An empirical study. European Journal of marketing, 335(6),470-487.

31 Chen, C & Y.Chang 2008. Airline brand equity, brand preference and purchase intentions- The moderating effects of switching costs. Journal of Air transport management, 14:40-42.

32 Chang, H.H & Y.M.Liu, 2009. The impact of brand equity on brand presence and purchase intention in the service industries. The service industries Journal, 29(12):1687-1706.

H3. Brand preference has direct and positive influence on consumer's purchase intention

H4. COO has moderating role in relationship between brand equity and purchase intention.

H5. COO has moderating role in relationship between brand preference and purchase intention

Roth and Romeo (1992): [33]According to them, the country of origin image scales measures the consumer's perception of the image of the country where the brand originates from there. Assuming that respondents have their idea about the country of brand which they are using now, they identified 4 basic dimensions which include innovativeness, design, workmanship, and prestige. **Chung** *et al.* **(2009) & Wong** *et al.* **(2008):** [34] expressed their idea about the importance of manufacturer country effects on young people in this way: "According to the globalization and since young people used to see the products from around the world which is produced by a country except the owner of that brand, therefore this issue confirm this claim that country image has no great effect on young people." These consumers feel the brands belong to countries with good image are more reliable rather than brands that produce in countries with a less desirable image.

Han (1989):[35] has attempted to explain two roles of Country image; one is the halo effect and the other is the summary effect. On one hand, when consumers are not familiar with a product or a brand, the consumers tends to rely on halo effects which can indirectly affect consumers' product/brand attitudes when inferring the product/brand attributes. On the other hand, they summarize their beliefs regarding product/brand attributes and this summary construct directly influences consumers' attitudes toward the product/brand, when they are familiar with the product/brand. **Thakor & Katsanis (1997):**[36] have developed a model of brand and country effects on quality dimensions. They suggest that country image cues affect quality perceptions both directly and through the brand cue. Conceptualization and measurement of "country image dimensions" appears to have evolved in a somewhat unsystematic manner in the COO literature.

33 Roth and Romeo (1992): Matching product category and country image perceptions: A framework for managing country of origin effects.Journal of international business studies, 23 (3):477-497.

34 Wong Y.C, M.J. Polonsky and R.Garma, 2008. The impact of consumer ethnocentrism and country of origin sub components on high involvement products on young Chinese consumers product assessments. Asia pacific Journal of Marketing, 20(4):455-478.

35 Han, C.M (1989): Country Image. Halo or summary construct? Journal of marketing research May, 26:222-29.

36 Thakor and Katsanis (1997): A model of brand and country effects on quality dimensions: Issues and implications. Journal of international consumer marketing, 9(3),79-100.

Aker (1991):[37] According to him, brand credibility is an impotent antecedent of quality perceptions that may also be influenced by country image in the interaction with a particular brand trait. Thus, when a brand can be distinctively differentiated from other brands on personality and its COO image is perceived as positive. In contrast, when brand personality is positive, while COO image is negative, the connection between these two variables would become "loose". He suggested 3 hypotheses:

H1: Country image moderates between the relationship between brand personality and brand credibility.

H : Ancient country image influences brand credibility more positively on sincere b²and than on exciting brand.

H : Modern country image influences brand credibility more positively on exciting b³and than on sincere brandmore easily influenced by their perceptions of that brand's personality.

Roth and Romeo (1992):[38] According to them, that country image appears to be a multidimensional construct. As such "Ancient" versus "modern" are placed at the two opposite extremes to illustrate a common belief that these 2 concepts cannot be integrated - a modern brand would not possess "stubborn" traditions. Based on the above definition, "Ancient" links with a country's history, heritage and cultural abundance while "Modern" relates to its technical advancements, inventiveness and technology innovation. An advantage for such a treatment is to magnify the ambiguity in the market dilemma and provide more useful insights for managerial decisions.

Wang X.H. & Yang Z.L. (2008):[39] In their Research, studied the fit and match of country image with the two most dominant brand personalities, extending the prior studies by giving a more in-depth understanding of how they differ and perform under different settings. It provides evidence to explain what exactly means a favorable or unfavorable country image and what gives a negative or positive COO effects. In this sense, an ancient country image could be concluded to be unfavorable and exert a negative COO effect to exciting brands. It is suggested that any county image may not have absolutely good or bad meaning in its nature, but the COO effect depends on its fit and match with the product category and the particular brand.

---

37 Aaker, D.A.(1991): Managing brand equity: Capitalizing on the value of a brand name. The fress press, NewYork, NY.

38 Roth and Romeo (1992): Matching product category and country image perceptions: a framework for managing country of origin effects. Journal of international business studies, Vol.23 No.3, pp.477-97.

39 Wang X.H & Yang Z.L. (2008): Does country of origin matter in the relationship between brand personality and purchase intention in emerging economies: Evidence from chinas auto industry. International Marketing review, 25(4), 458-474.

Ravi P *et al.* **(2007):** [40]According to him, marketers need to go a step further and manage the country image, while also managing their marketing mix. In other words, apart from composing both good brand strategies to differentiate themselves in the market, companies need to either stress or downplay the COO effect, depending on whether the relevant COO image is positive or not, in order to achieve market success. **Bannister and Saunders's (1978):** [41]According to them, image is created by variables such as representative products, economic and political maturity, historical events and relationships, traditions, industrialization, the degree of technological virtuosity and etc., which will have effects upon consumer attitudes additional to those emanating from the significant elements of the products.

Keller, K.L (1993):[42] According to him, Country-of-origin (COO) can be thought of as a kind of branding strategy, since both attempt to develop a competitive advantage based on familiarity with either the brand name or the country-of-origin. Sometimes the country-of-origin is used in addition to a brand name, and this is what referred to as a secondary association is. At other times the country-of-origin is used instead of a brand name (i.e. as *the* brand name). The benefit of using a country-of-origin strategy rather than a "traditional" branding strategy is the *added dimension of a country image* since people often already have a relationship to, or opinions about, different countries. By spinning on these relationships it is possible to sell products and services that would have little opportunity of making it internationally on their own. The COO effect depends, among other things, upon *the knowledge of the COO* in the target market

Papadopolos (1993):[43] According to him, the higher the level of market globalization, the greater the potential of COO in influencing consumer behavior due to increased country exposure in the media, greater awareness and growing presence of foreign products, and serious benefits of simplified information processing with increased complexity in markets and products. Information on the origin is well suited for these purposes because it can be used as an indicator of product quality and acceptability. **Kotler, Haider, & Rein, (1993):**[44] defined an image of a place as "the sum of all those emotional and aesthetic qualities such as experience, beliefs, ideas, recollections and impressions that a person has of a place". In this definition it is evident that an individual forms an image based on her personal frame of reference.

---

40 Rati P *et al.* (2007): Country image and consumer based brand equity: relationships and implications for international marketing. Journal of international business studies, 38,726-745.

41 Bannister and Saunders (1978): UK consumers attitudes towards imports; The measurement of national stereotype image, European journal of marketing, 12 (8),562-570.

42 Keller, K.L 91993): Conceptualizing, measuring and managing customer based brand equity, Journal of marketing, Vol.57,pp,1-22.

43 Papadopolos (1993): Product-Country Images-Impact and Role in International Marketing", *International Business Press,* New York, 1993

44 Kotler, Haider and Rein (1993): Marketing places, The fress press, New York, 1993.

Gunn (1972):[45] According to him, images of places are formed on two levels and referred to as "organic" and "induced". The organic image of an area is created from general exposure to information such as newspaper reports, magazine articles, television reports etc. This process of image formation can be seen as a result of lifelong socialization. In marketing the focus is on the second level which could be advertising attempts to "induce" an image into the consumer's evoked set, and thereby increase the probability of a purchase decision in favor of a product.

Chon (1990):[46] According to him, a moment of *primary image,* this reflects a relationship between the consumers' needs and expectations, and the ability of the image to match these. The marketer seeks to activate specific associations from the CI, and to match these with important characteristics in the target market through the design of all the marketing mix components. If all these components successfully come together, it is likely that an intended image association will be established. It is only when this succeeds, i.e. when the image creating moment occurs, that the CI can develop and eventually become stronger and more multiplex.

Graby, f (1993): [47] identified the 'identity prism' of the country (like the concept of corporate identity) consists of physical (geography, natural sources, demography), cultural (history, culture), personal (name, flag, celebrities), relational (with governments, international organizations) and controlled (conscious formation of country image) elements". Country image comprises many elements: national symbols, colors, clothing, typical buildings, objects, tunes, pieces of literature, specialties of the political system, customs, historical heritage etc.

Jenes, B(2007):[48] ascribes a so-called 'umbrella function' to country image, as its elements are made up of the totality of the country's specific products, brands and various organizations and their images. According to the second approach, the country itself is a complex product, made up of a large number of elements. (Thus country image is considered a normal product image, yet with more diverse, complex and complicated characteristics). Regarding its direction, the country image can be internal image (self image) and external image (mirror image), similarly to the classification of product image.

Malota's (2001):[49] According to him, there is a relationship between the respondent's self-confidence and their view on a country image. Regarding this

45 Gunn (1972): Vacationscape: Designing tourist regions, Austin: University of Texas Press.

46 Chon, K.S., (1990), "Tourism Destination Image Modification Process and Marketing Implications", *Tourism Management,* March 1991, pp. 68-72

47 Graby, F. (1993): Countries as corporate entities in international markets, in Product-Country images. Impact and role in international marketing.

48 Jenes, B (2007): Connection between ecologically oriented consumer behavior and country image. Marketing and menedzsment, 2007/6.pp-34-43.

49 Malota, E. (2001): Consumer ethnocentrism: the effect of stereotypes, ethnocentrism and country of origin image on the evaluation of domestic and foreign products. Ph.D. thesis BUESPA

assumption the level of self-confidence was also measured on a five-point scale, with the following categories: far below average, somewhat below average, average, somewhat above average and far above average. Here we found four significant relationships. The higher the level of people's self-confidence the more they agreed to three of the statements (Successful country, Country with great culture, Decent, clean country), while one statement (Country of fair and honest people) yielded fluctuating results depending on the level of self-confidence.

Verlegh & Steenkamp (1999):[50] According to them, there are dimensions that clearly reflect consumers' cognitive perceptions about a country, past research has shown that *"country of origin is not merely a cognitive cue for product quality, but also relates to emotions, identity, pride and autobiographical memories"*. A *cognitive* component, which includes consumers' beliefs about the country's industrial, technological as well as political background; an *affective* component that describes the country's symbolic and emotional value to the consumer, and a *conative* component, capturing consumers' desired interaction with the sourcing country. Since cognitive, affective and normative processes are interacting in consumer decision-making,

Papadopoulous (2000):[51] The main practical contribution of his research lies in the explanation of a greater proportion of variance in product image and buying intention, thus providing international marketers with clear "do's" and "don'ts" when operating in foreign markets. From a theoretical perspective, linking consumer ethnocentrism, national identity and consumer cosmopolitanism to established constructs such as country image and product image will contribute to the theoretical explanation of the formation of such images and how they impact on purchase intentions.

Schooler, R.D (1965):[52] According to him, Country-of-origin image (COI) is an important driver of consumers' evaluation of products originating from different countries. He is generally considered to be the first researcher to empirically study this effect. He found out that products, identical in every respect except for their country of origin, were evaluated differently by consumers. **Knight and Calantone (2000):** [53] lament that *"despite hundreds of studies on the COI effect, little is known about the cognitive processing that occurs during COI-based product evaluations."* It is this gap in the literature that the present study seeks to address, focusing specifically on the role of potential antecedents of country image perceptions as well as their impact on product image, product evaluations and purchase intentions.

---

50 Verlegh, P.W.J & Steenkemp, J-B.E.M (1999): A review and meta analysis of country of origin research. Journal of economic psychology, 20 95),521-546.

51 Papadopoulos, N., Heslop, L. A., & IKON Research Group (2000). A Cross-national and Longitudinal Study of Product-Country Images with a Focus on the U.S. and Japan. *Working Paper, Report No. 00-106,* Marketing Science Institute: Cambridge.

52 Schooler, R.D (1965): Product bias in Central American common market. Journal of marketing research, 294), 394-397.

53 Knight, G.A & Calantone, R.J. (2000). A flexible model of consumer country of origin perceptions. International marketing review, 17(2).127-145.

Turney (2000):[54] According to him, terms "good image" and "bad image" simply reflect how positively or how negatively people who are exposed to the image respond to the person or organization represented by the image. A country image is defined 'as the total of all descriptive, inferential and informational beliefs one has about a particular country'. An image of a place is defined as 'the sum of all those emotional and aesthetic qualities such as experience, beliefs, ideas, recollections and impressions that a person has of a place'. **D'Astous and Boujbel's (2007):**[55] have developed a country personality scale that profiles a particular country along six personality dimensions. They specifically stress that country personality "represents *one particular way* of looking at country images and should be considered as a *complement* rather than as a substitute to existing country image measuring instruments". In addition, the relationship between "traditional" attribute-based conceptualizations of country image and country personality has not been assessed so far, leaving it unclear whether future COO research should focus on country image, country personality, or both constructs at the same time

Yoo and Donthu, (2001):[56] According to them, marketing literature does not explain whether consumer-based equity of a brand is linked to the macro and micro images of the country in which it is produced. Further, it is not clear whether the impact of country image on the consumer-based equity of a brand is product category specific. Therefore this study examines whether macro country image, micro country image and consumer-based brand equity are related, and whether these relationships hold across different product categories. **Martin and Eroglu (1993):**[57] believed country image as 'the total of all descriptive, inferential and informational beliefs one has about a particular country'. This macro country image is argued to be different from a consumer's attitude towards products from a given country. Martin and Eroglu proposed that country image has three underlying dimensions, namely economic, political and technological.

Kim and Chung (1997):[58] According to them, there is a link between country and brand images. These researchers have argued that consumers have country-specific brand images. Further, a country's image in a given market might be affected by the performances of major brands originating from that particular country. Thus the marketing literature suggests a bidirectional relationship between country

54 Turney, Micheal Images can be natural or constructed perceptions, On - line Readings in PR (http://www.nku.edu/~turney/prclass/readings/image.html) 2000.

55 D'Astous, A. & Boujbel, L. (2007). Positioning Countries on Personality Dimensions: Scale Development and Implications for Country Marketing. *Journal of Business Research*, 60(3), 231-239.

56 Yoo & Donthu, N (2001): Developing and validating a multidimensional consumer-based brand equity scale', Journal of Business Research 52(1): 1–14.

57 Martin, M. and Eroglu, S. (1993) 'Measuring a multi-dimensional construct: country image', Journal of Business Research 28(3): 191–210.

58 Kim C.K. and Chung, J.Y. (1997) 'Brand popularity, country image and market share: an empirical study', Journal of International Business Studies 28(2): 361–387.

image and brand image. Given that a brand image is a set of consumer brand associations organized in a meaningful way the literature would appear to suggest a relationship between country image and consumers' brand associations.

Agarwal and Sikri (1996):[59] observed that consumer's country image beliefs in relation to a familiar product category transferred to new products offered from the same country. They argued that the country-of-origin cue operated in a manner similar to brand name in the transference of beliefs. Thus such transference of beliefs may extend to brands, and consumers could be more loyal to brands made in countries with favorable images. **Zeugner-Roth** *et al.*, **(2008)**:[60] adopted the constructs of Country Brand Equity (CBE) and linked them to Country Image (COI). They found that country based brand equity is influenced by country image perceptions and that country brand equity positively impacts on product preferences. They also discerned that product preferences are not directly influenced by country image perceptions.

Jaffe and Nebenzahl (2006):[61] In their hypothesis, point out that,, *a highly positive country image as a source of the products will add to the positive image of the brand and as a result to its equity and value"*. It could be argued that image is a much larger issue and encompasses, cultural dimensions as well the imagery of a country. Equity could also be interpreted in different ways. Nonetheless the idea of separating equity and image is interesting and warrants further study. **Wall and Heslop (1991):**[62] proposed that purchasing foreign-made products may be seen as immoral and unpatriotic because it has an adverse impact on the domestic economy; hence, consumers tend to purchase local products even if the quality is lower than that of imports.

O'Shaughnessy, J., & O'Shaughnessy, N. (2000). In their study, country brand equity turned out to be an important driver of consumers' product preferences in all situations explored. Also, the role of country image as antecedent of country brand equity was supported by the research. However, they could not detect a significant relationship between country image and product preferences. Country brand equity is therefore an important intervening variable between consumers' perceptions about a country (i.e. country image) and preferences towards products from that country. **Jaffe and Nebenzahl (2001):** [63]According to them, four possible

59 Agarwal S. and Sikri, S. (1996) 'Country image: consumer evaluation of product category extensions', International Marketing Review 13(4): 23–39.

60 Zeugner- Roth, K.P.Diamantopoulous, A., Montesinos. M.A. (2008) Home Country Image, Country Brand Equity and Consumers® Product Preferences. *Management International Review*. 48(5). 578-602.

61 Jaffe, E.D and Nebenzahl I.D. (2006) *National Image and Competitive Advantage: The theory and Practice of Place Branding*. Copenhagen: Copenhagen Business School Press

62 Wall & Heslop(1991): Impact of country of origin cues on consumer judgments in multi cue situations: a covariance analysis. Journal of Academy of marketing science.19.105-113.

63 Jaffe, Eugene D- NEBENZHAL, Israel D.2001: National image and competitive

strategies can be used. The first scenario considers those companies that have both a strong country and brand image. The second scenario looks at companies with a weak country image but strong brand image. The third possibility covers companies with weak brand image, but with strong country image, and the fourth version is companies with both weak country and brand image.

Lee & Simon (2006):[64] In their hypothesis analyzed how consumers' perceptions on the quality of products are influenced by the marketing appeals of multinational firms and by the country of origin effects. Their findings suggested that the attitudes of consumers towards country of origin and corporate image exert a great deal of influence on their perceptions of product quality and purchase behavior, the effect of certain country image appeals on the purchase behavior and moderated by socio-economic and national cultural characteristics.

Author	In their thesis
Bannister- Saunders (1978)[1]	Country image is an overall image that is constituted by variables like peculiar products, economic and political development, historical events and relationships, traditions, level of industrialization and of technical development
Desborde (1990)[2]	Country of origin image is an overall image of country in consumer's minds. It reflects a country's culture, political system and its level of economic and technological development.
Martin-Eroglu (1993)[3]	Country image is a set of normative inferred and informational beliefs of individuals on a country.
Kotler *et al.* (1993)[4]	Country image is a sum of people's beliefs, ideas and impressions about a certain country.
Szeles (1998)[5]	Country image is an external and internal framework of the opinions and beliefs on a people, nation and country and the simultaneous objective and subjective psychological contents of heterogeneous and generalized value judgment thereof.
Allred *et al.* (1999)[6]	Perceptions and impressions that institutions and consumers have of a country. This prior impression is based on the economic state of the country, its political structure, its potential conflicts with other countries, its labor market conditions and other environmental Factors.
Verlegh – Steenkamp (1999)[7]	A mental interpretation of a country's inhabitants, products, culture and national symbols.

advantage- The theory and practice of country of origin effect. Copenhagen business school press, Handelshojskolens Forlag.

64 Lee and Simon (2006): Cultural Variation in Country of Origin Effects. Journal of Marketing Research, vol. 37 (3), pp. 309-317

Avraham-Ketter (2006)[8]	Country image is constituted of several elements, among others of the country's location, political structure, economic situation, the stability of its government etc. Even though this image seems dynamic, it is based upon stereotypes.
Brijis *et al.* (2011)[9]	Country image represents all that a consumer attaches to a country and its inhabitants (and not its products)

2.3. Trade Regulations and Plantation Commodities

India's signing the treaty with ASEAN would eventually eliminate duty on 80 per cent of the goods traded at present. The applied Most Favored Nation (MFN) tariff rates on products such as coffee, palm oil, pepper and tea will be brought down in phases and in the case of tea, the reduction commitment is 100 to 50 per cent by 2019. There are apprehensions that reduction in import tariff on tea will affect Indian tea industry adversely, and through the deficient Rules of Origin China's cheaper tea may enter Indian markets through ASEAN. Indonesia and Vietnam are major tea exporting countries among ASEAN countries (B H Nagoor and C Nalin Kumar). It would be great advantage for Indonesia and Vietnam to explore the huge Indian domestic market under the Free Trade Agreement (FTA). The reduction in import tariffs on tea from ASEAN countries, there are apprehensions that, ASEAN countries tea exports to India may increase drastically and will affect Indian tea industry adversely. The results of our preliminary analysis show that the export unit value of tea of Thailand, Vietnam and Indonesia is lower than the import unit value of Indian tea. So, at present India is importing tea at a higher rate. Due to export price advantage, Thailand, Vietnam and Indonesia may increase their tea exports to India. For instance, Vietnam tea has a price advantage of 26.84 per cent over India.

Roopam Singh and Ranja Sengupta (2009)[65] said in their paper that EU-27 has an advantage in poultry, dairy and dairy products, cereals other than rice (basmati and non-basmati), fruits and vegetables, coffee, mate, tea, sugar and olive oil over India. The EU India FTA may lead to increased imports of the products, especially more value added products, in which EU has an advantage. This will affect dairy, coffee, mate, tea and wheat farmers, and fruits and vegetable growers and local agro based processing industries.

2.4. Food Labelling

The information presented on a package of a product is considered to be very influential to the consumers' buying process. By paying attention to the food label information, consumers can ensure that they and their families eat the correct amounts of nutrients. They can also avoid overeating and keep allergens away from themselves and their families. Labels also assist consumers in choosing

65 Roopam Singh and Ranja Sengupta (2009): The EU India FTA in Agriculture and Likely Impact on Indian Women, http://in.boell.org/sites/default/files/downloads/Summary_agr.pdf

products, which are manufactured in a way that complies with their moral standards in terms of matters such as environmental sustainability and fair trade. Also, labels can provide certification to the consumer of authenticity which may be of value to some. For food manufacturers, labels are a versatile tool for communicating information regarding nutrition, qualities of the product, process-related characteristics and other relevant information, that marketers find relevant in order to promote and position their product.

Food labels are believed to have the ability to have an effect on food choices and dietary behavior (Mackison, Wrieden, & Anderson, 2010) and are commonly acknowledged to have a central role in communicating product-related information to consumers. Food labelling has become an important policy tool to enable consumers to get thorough information on the contents and the composition of food products. Certain mandatory regulations are being developed in the European Union **(Campos** *et al.*, **2011)**[66], but so far food labelling on pre-packaged foods is a voluntary scheme, except in the case of certain health and quality claims.

Joseph *et al.* (2009):[67] developed a conceptual model of heterogeneous consumers that examines the consequences of partial Mandatory Country of Origin Labelling (MCOOL) implementation on welfare and diversion. Numerical simulation results showed that diversion is possible in the partial MCOOL scenario and the higher the perceived quality of domestic fish the greater the diversion of low-quality imports to the non-labeled market. Real consumer surplus was greatest under total MCOOL implementation when quality differences between domestic and foreign fish are perceived to be great. Results indicated that the majority of respondents read the information provided on food labels. In addition to that, more than half of them would like to see labeled food items on stores◎ shelves. Among the attributes written on the labels, the three most important that were checked by respondents were expiration date, list of ingredients and the country of origin.

Caswell and Padberg (1992):[68] According to them, food labels play important roles in the food marketing via their impact on product design, advertising, consumer confidence in food quality, and consumer education on diet and health. **Mackison, Wrieden, & Anderson, (2010):**[69] acknowledged having a central role in communicating product-related information to consumers. Food labelling has become an important policy tool to enable consumers to get thorough information on the contents and the composition of food products.

---

66 Campos, S., Doxey, J., & Hammond, D. (2011). Nutrition labels on pre-packaged foods. A systematic review. Public Health Nutrition, 14(8), 1496-1506.

67 Joseph *et al.* (2009): Joint FAOWHOCAC (2009). Food labeling: Food & Agriculture Org. www.fao.org.

68 Caswell JA, Padberg DI (1992). Toward a more comprehensive theory of food labels. Am. J. Agric. Econ. 74(2):460-468.

69 Mackison, Wrieden and Anderson (2010): Validity and reliability testing of a short questionnaire developed to assess consumers' use, understanding and perception of food labels. *European Journal of Clinical Nutrition, 64*(2), 210-217.

Hieke, S., & Taylor, C. R. (2012):[70] According to them, there exist two main aspects that literature focuses on in this regard; the analysis of socio-demographic factors and the other personal factors related with the use and comprehension of food labels. Female users, higher educated consumers, persons with high income and certain age-groups have generally been found to use food labels more frequently and understand them better. There seems to be some indications that women are more likely than men to use food label information, even though research has brought mixed results. Age seems to be negatively related to comprehension of food labels, while it does not seem to have any strong relation to the use of food labels.

Grunert and Wills (2007):[71] have developed a framework that reflects consumers' decision-making process relating to their understanding and use of food labels. The framework has been developed based on two streams of research; 'consumer-decision making' and 'attitude formation and change'. According to them the likelihood of exposure increases if consumers actually search for the label information, even though active search it is not a precondition for exposure, as it can also be completely random.

Mc Cullough J., and Best, R (1980):[72] According to them, most legal regulations concerning food product labelling are conceived and implemented on the basic premise that dissemination of information in greater quantities and details will facilitate consumers in making better brand choice decision. This basic premise is in turn based on the assumption that consumers are aware of the information being provided on the product label. Further, policy interventions regarding mandatory disclosure of food product information also assume that consumers can comprehend and interpret the information on the food label.

Wandel (1995):[73] According to him, women, the highly educated and those who are on special diets, tend to read the food labels to a greater extent than others. Many investigators have also found that the interest in reading the food labels increases with age up to the mid-fifties, and thereafter it declines. The typical food label reader is reported to be a middle-aged woman with high education.

Guthrie *et al.* (1995):[74] According to him, Indian consumers assign very high

70 Hieke, S& Taylor C. R. (2012). A critical review of the literature on nutritional labeling. *Journal of Consumer Affairs, 46*(1), 120-156.

71 Grunert and Wills J. M. & Fernández-Celemín, L. (2010) Nutrition knowledge, and use and understanding of nutrition information on food labels among consumers in the UK. *Appetite*, 55, 177-189

72 Mc Cullough J., and Best, R., 1980: Consumer Preferences for Food label Information: A Basis for Segmentation. *Journal of Consumer Affairs*, 14(1): 180-182.

73 Wandel M., 1997:Food Labeling from a Consumer Perspective. *British Food Journal*, 99(6): 212-214.

74 Guthrie, J.F. Cleveland, L.E., Welsh, S., 1995:Who uses nutrition labeling, and what effects does label use have on diet quality? *Journal of Nutrition Education*, 27 (4), 163–172.

importance to information about food ingredients and nutritional contents of the food. However, as compared to these aspects of food labels, information on serving size and short phrases has lower priority among these consumers. They also have very strong preference for brand and the taste of the product. This preference makes them purchase a packaged food item even though it may not meet their criteria of healthy food or may contain some harmful ingredients. The results give a clear indication that label information is generally gender and age insensitive though its use assumes significance with the income levels, education and occupation of the consumers. The awareness regarding label information is dependent on the consumer's level of education, income and kind of occupation. He concluded the outcomes of the study that reveal the level of awareness about food labels and their usability among Indian consumers of packaged food products.

The **Special Eurobarometer on Europeans' attitude towards food security, food quality and the countryside** found that, on average, "a substantial majority (71%) [of EU citizens] say that the origin of food is important" (as in BEUC's study, this proportion was lower (57%) among younger people who said that origin was important to them). Looking at specific country results, citizens in Austria, France, Poland and Sweden were respectively 78%, 75%, 71% and 79% to reply that origin is important to them when buying food. These figures are similar to that observed in BEUC's research.

The **European Commission's study of the functioning of the meat market for consumers in the EU**[75] addressed the question of consumers' motivation and priorities when shopping for meat. Whether the meat comes from their country is the fifth factor European consumers say they take the most into account, behind the freshness, taste, hygienic aspect and price of meat. Factors they look at the most when buying fresh meat are the price per kilogram, the price, the use-by/best-before date and the country of origin, followed by the producer, origin certifications, the list of ingredients and animal welfare labels. For meat products, consumers primarily pay attention to the use-by/best-before date, the price, the price per kilogram and the country of origin, followed by the producer, the list of ingredients, origin certifications and the nutritional value. The findings of this study are in line with that of BEUC's survey and confirm consumers' interest in the origin of meat and processed meat products.

Most recently, in July 2012, the **Czech Association of Consumers DTEST** carried out a survey[76] on origin labelling using the questionnaire drafted for the purpose of BEUC's research, which was administered online to a panel of subscribers and followers. Although the panel of respondents was not representative of the average population in the Czech Republic, the data collected still provides valuable insight into Czech consumers' interest in, and understanding of origin labelling.

75 To be published. Results available at: http://www.llkc.lv/upload_file/401466/Paulina%20Gbur_Outcomes_Kaunas%20[Read-Only]%20[Compatibility%20Mode].pdf

76 The survey was conducted online between 13-20 July 2012 on a panel of circa 1700 subscribers and followers of DTEST

89% of respondents replied that origin is important to them when buying food (as compared to 80% of the representative panel of Czech citizens interviewed in the Eurobarometer survey), essentially because it helps them assess the safety and quality of food.

A majority of Czech consumers (71%) want to know the country their food comes from and, when asked about their preferences for origin labelling on processed foods, 83% reply they are equally interested in knowing the country of farming (of animals or fruits/vegetables) as that of processing.

The **Greek consumer organization KEP KA** conducted similar research in September 2012 (also based on BEUC's questionnaire)[77]. The panel of respondents was not representative of the Greek population as the questionnaire was published on KEPKA's website; however, the survey results confirm the findings of the Eurobarometer report No 389. 92% of Greek consumers surveyed declared that origin is important to them when buying food (as compared to 90% of Greek citizens in the Eurobarometer report). Origin comes in 3rd place in terms of the factors they say they look at the most, behind taste (98%) and price (97%) but before appearance (63%), convenience (61%) and brand (49%). Almost all respondents (97-98%) find it important that origin is labeled on meat, fish, milk and dairy products, and fresh fruit and vegetables. A vast majority of them also want origin labelling on processed fruit and vegetables (91%), staple foodstuffs (86%) and coffee/tea (76%). Greek consumers mostly relate the origin of food to safety (71%), quality (61%) and ethical (52%) concerns. As for the geographical level they are interested in, 60% of Greek consumers want to know the country their food comes from or even, for 37% of them, the specific region. Only very few respondents (3%) replied they would be satisfied with EU/non-EU origin indication.

These outcomes are helpful for policy makers as well as food companies in designing appropriate strategies for improving awareness among consumers and ensuring that their usability is improved. The outcomes also help the food companies in deciding which type of information on the food labels matters the most to the aware consumers in making rational food choices.

The review of literature revealed that there are enough studies globally done related to promoting products on COOL. Based on the review, it can be summarized that the variables such as evaluation criteria, attitude towards foreign brands, consumer ethnocentrism and perceived country image have a greater impact on purchase intention.

---

77 1556 Greek consumers participated in the survey. The survey was conducted in September 2012, using a questionnaire published on KEPKA's website (98% of respondents). At the same time, a printed version was disseminated through KEPKA's offices to individuals not having access to the internet (2% of respondents). The full results of the survey are available on KEPKA's website at: http://kepka.org/index.php?option=com_content&task=view&id=1923&Itemid=237

Research Design

3.1. Need for the Study

Globalization of markets presents considerable challenges and opportunities for domestic and international marketers. With a current border-thinning global economy, consumers around the world have been increasingly exposed to foreign products, giving them more buying choices. The demand supply situations for the selected plantation commodities such as tea and coffee in the Asian continent have undergone a rapid transformation due to the growth of the world economy and lowering of trade barriers. The tariff reduction lead to a significant increase in India's imports from the ASEAN countries. The surge in imports will have an adverse effect on exploiting the Indian consumers and intrude into the domestic market. To give a barrier to such a situation, in the liberalized global market, it is high time to promote these commodities with an innovative branding strategy.

The project aims to evaluate the level of consumer ethnocentrism and to assess the Consumer Ethnocentric Tendencies and its implications on the attitude of Indians towards consumption of domestic vis-à-vis foreign goods so as to promote using 'country of origin labelling'.

Although there are several studies on consumer ethnocentrism across different countries, there is a dearth of literary on any such studies conducted in India in general and for plantation products in particular. The attitude of consumers towards foreign goods and their preference over domestic goods is an important piece of information for the marketers which will help in branding and promotion. Though there are studies done in this area, there is not much study pertaining to coffee/tea. Moreover, the dynamics of consumer behavior is such that it changes very fast and hence necessitates longitudinal studies.

Researchers have come out with findings that consumers' country images can affect the equity they associate with a brand from that country. Studies also proved that country image can influence the key dimensions of brand equity such as brand associations, perceived quality and brand loyalty. However, there is dearth of organized studies to find out the perceived image of Indians on their own country. Several studies proved that these variables heavily influences the purchase intention of the consumer, however they are least understood. The above said reasons justify the need for this study.

3.2. Scope of the Study

The aim of the study is to find out the feasibility of promoting coffee/tea as "Indian Coffee/tea' in the domestic market. The scope of the study is limited to the following broad research variables:

- Awareness level of Indian consumers on coffee/tea producing countries, quality and brands.
- Criteria in evaluation of coffee/tea brand and the relevance of country of origin
- Attitude towards foreign brands
- Consumer Ethnocentrism and the Country Image.

The study focuses only coffee and tea as plantation products. The study area is limited to Karnataka state.

3.3. Objectives of the Study

The general objective of the study is to find out the effect of country of origin labelling on the purchase intention of Indian consumers for the selected plantation products tea and coffee.

Following are the specific objectives:

- To explore the level of awareness among the consumers of coffee and tea regarding the places where these commodities are produced and the quality.
- To determine the extent of 'country of origin' effect compared to other aspects upon product evaluation and decision making.
- To measure the attitude of consumers towards foreign brands of coffee and tea
- To assess the Indian consumers ethnocentrism pertaining to the purchase of tea and coffee.
- To examine the perceived image of India as a country among the domestic consumers.

3.4. Work plan and its Phasing

The Work Plan, different phases and the time spent is shown in the table below:

Tab: 3.1: The Work Plan and Phases

Work Plan	Month											
	M1	M2	M3	M4	M5	M6	M7	M8	M9	M10	M11	M12
Review of Literature & Finalization of the Hypotheses												
Finalization of the tool for data collection (questionnaire)												
Pilot Study												
Field Work												
Data entry & analysis												
Report Writing												

3.5. Research Methodology

Research Method

The study is exploratory and descriptive in nature. A survey method has been deployed. The survey has been done across various districts of Karnataka state and the scope of the study limited to tea and coffee. Online survey has been commissioned for the employed professionals in Bangalore city. Mostly, a field survey has been carried out where, a group of respondents were asked to assemble and through a focus group discussion, the purpose and the context of the survey are briefed.

Sources of Data

Both primary and secondary data were collected for the study. Data related to the coffee and tea industry in India and markets for plantation products were collected through secondary sources such as commodity boards, books and other published materials. Secondary data were also collected relating to the research variables and the collected literature were compiled and analyzed to finalize the hypotheses and decide the research focus.

Primary data were collected from individual consumers, experts from industry and the stakeholders of coffee/tea industry. Primary data from individual consumers were collected to assess the consumers' awareness level, ethnocentrism, extent of 'country of origin' effect upon product evaluation and decision making and perceived country image through a sample survey among the consumers.

Research Variables and Tools for data collection:

Based on the literature review the following variables found to be relevant and considered for further study:

- Awareness level of Indian consumers on coffee/tea producing countries, quality and brands.
- Criteria in evaluation of coffee/tea brand and the importance of country of origin in brand evaluation.
- Attitude towards foreign brands
- Consumer Ethnocentrism and the Country Image

A structured questionnaire is designed and administered among the target respondents as a tool for data collection. The questionnaire contained sets of questions pertaining to the research variables. Copy of the questionnaire is attached as Annexure – I. Based on the review of literature, a conceptual framework has been developed by the researcher and the same is depicted in the figure 3.1. The research variables have an influence on the purchase intention through changing the consumers' attitude towards domestic brands. The relationship between research variables and attitude towards domestic brand is shown in fig 3.2.

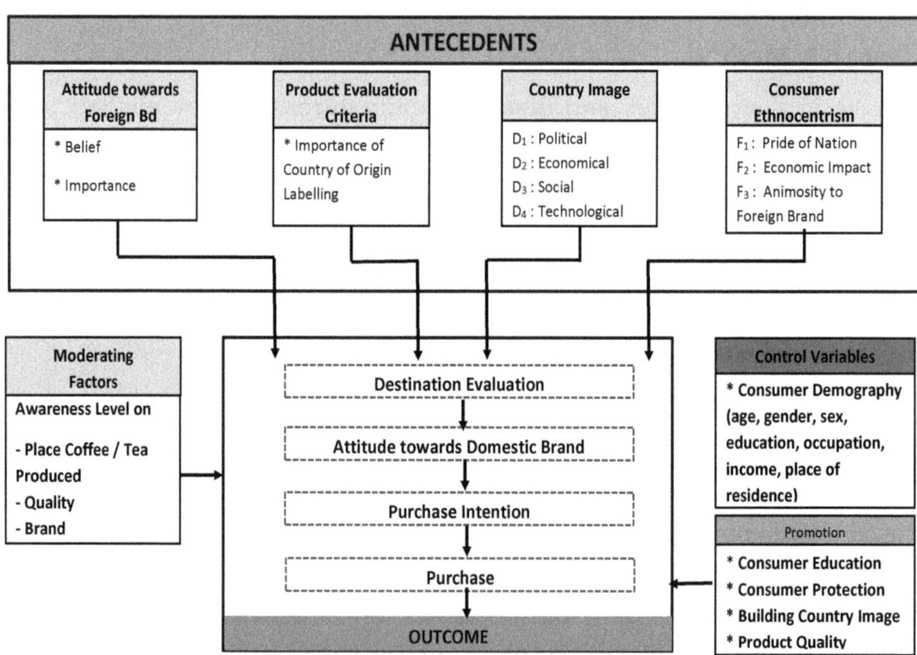

Fig. 3.1: Research Variable and Framework

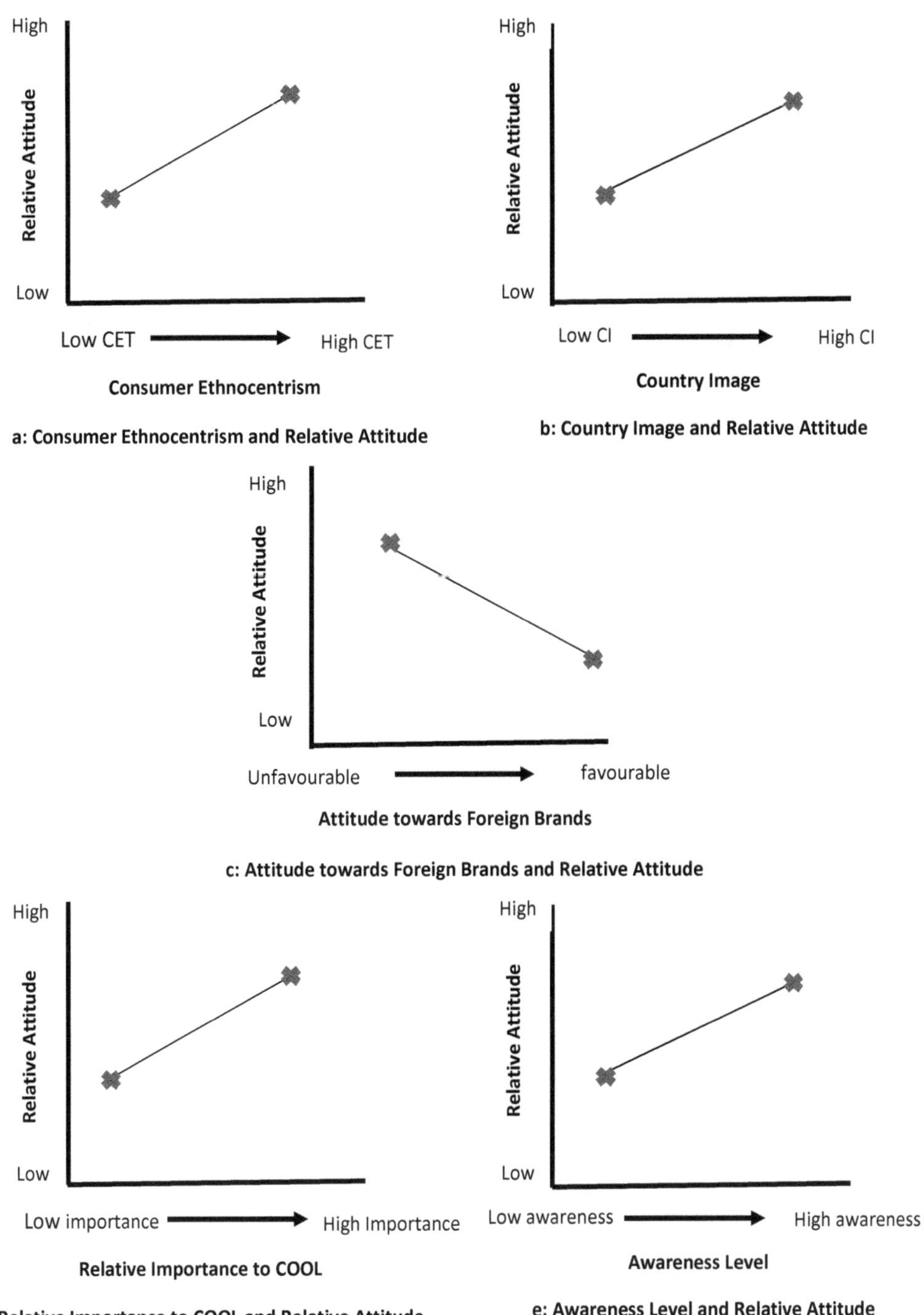

Fig. 3.2: Effect of Research variables and Consumer Attitude towards Domestic Brand

Awareness level: The consumers' awareness level on the places were coffee/tea grown, quality of these products produced in various countries and the brands available are collected by questioning them on these issues and choose options from no awareness, little awareness and complete awareness.

Evaluation criteria: The importance of origin information to consumers when they buy coffee/tea and the extent of 'country of origin' effect compared to other aspects upon product evaluation and decision making is measured by asking the respondents to rate the various criteria such as aroma, availability, brand, country of origin, cuppage, ease of use, information label, package, price and taste. The respondents were asked to rate these criteria in a five point scale from low important to high important.

Attitude towards foreign brands: The Belief Importance model (B-I model) has been used to measure the attitude of consumers towards foreign brands. The B-I model allows the comparison of affective responses toward competing brands and the formula adopted is given below:

$$A_o = \sum_{i=1}^{m} B_{io} I_i$$

where,

A_o = Attitude toward foreign brand

B_{io} = Belief that foreign brand does well or poorly when its attribute (i) is compared with those of domestic brands

I_i = Importance of attribute (i) in selecting the brand

i = attribute 1, 2, ... m

Five attributes were identified and two sets of statements, one for belief and another for importance were developed and the respondents were asked to choose from strongly disagree to strongly agree in a five point scale.

Consumer Ethnocentrism: Consumer ethnocentric tendency scale (CETSCALE) developed by Shimp & Sharma (1987), was used to measure the respondents' ethnocentric tendency with some modifications to fit the Czech context. This scale has been validated by different studies in different countries like USA, France, Japan, West Germany, Korea, and Poland etc. In all the previous studies it has reported a Cronbach's alpha value of over 0.90 as an indicator of the reliability of the scale (Netemeyer *et al.*, 1991; Sharma *et al.*, 1995;Kaynak and Kara, 2002)

Country Image: Gallup (2000) scale was adopted to measure the country image. The scale measures the evaluation of a given country (internal image) with a 24-item 4-point scale ("strongly disagree", "disagree", "agree", "strongly agree").

A copy of the questionnaire is attached as **Appenxix I.**

Sample Design

Samples were chosen from entire Karnataka state and the sampling unit is individual consumer. The four revenue districts were identified such as Bengaluru, Mysuru, Belagavi and Gulbarga. The districts covered under each revenue division is shown in the table below. Samples were drawn from each district representing different categories such as: Student, Home maker, Agri. and allied activities, Salaried/ Entrepreneurs/ Self Business and Others.

Table 3.2: Sample Size Distribution

Sl. No	Name of the Revenue Division	Districts under Jurisdiction	Sample Size
1	Bengaluru	Bengaluru Urban, Bengaluru Rural, Chikkaballapur, Chitradurga, Davanagere, Kolar, Ramanagara, Shivamogga, Tumakuru	375
2	Mysuru	Chamarajanagar, Chikkamagaluru, Dakshina Kannada, Hassan, Kodagu, Mandya, Mysore, Udupi	375
3	Belagavi	Bagalkot, Belgaum, Vijayapura, Dharwad, Gadag, Haveri, Uttara Kannada	375
4	Gulbarga	Bellary, Bidar, Gulbarga, Koppal, Raichur, Yadgir	375
		Total Sample	**1500**

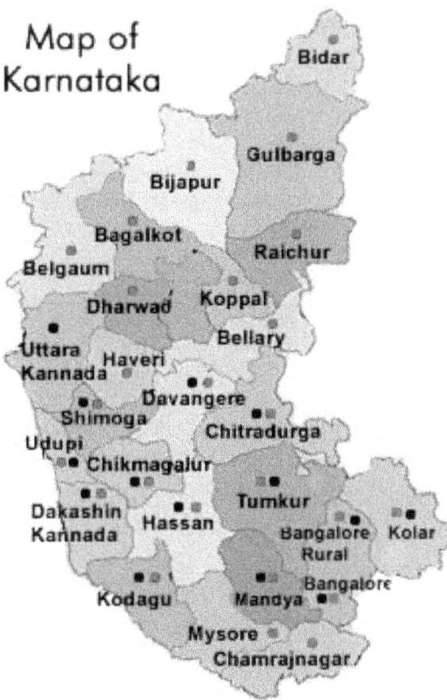

Fig. 3.3: Map of the Study Area

Tools for data collection

Tools for Analysis

The collected data were analyzed using Statistical Package and the data were compared with the demographic variables of the respondents and analyzed. The statistical analysis such as ANOVA, t-test, f-test, factor analysis were deployed to bring inferences.

Data Presentation and Analyses

4.1. Profile of Respondents

A field survey has been carried out in four regions i.e. Bengaluru, Mysore, Belgaum and Gulbarga of Karnataka district. The following section one by one will explain the demographic profile of the respondents. Table 4.1 shows the distribution of respondents across the various divisions.

Table 4.1: Distribution of respondents across the divisions

Division	Number of respondents		
	Total collected	**Rejected**	**Accepted**
Bengaluru	584	13	571
Mysuru	292	10	282
Belagavi	295	24	271
Gulbarga	332	20	312
Total	1503	67	**1436**

Source: Survey data

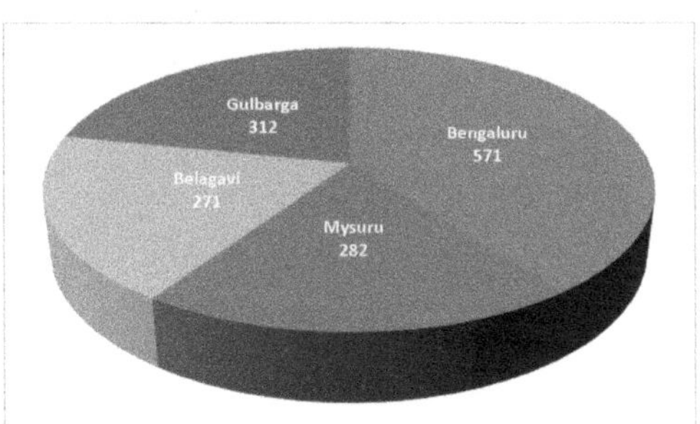

Fig. 4.1 Sample Distribution Across Division

Source: Survey data

Responses were collected from different categories across the various divisions of Karnataka state. Some of the respondents did not complete the entire questions in the questionnaire for various reasons. Some, after answering one part discontinued, some skipped few questions and yet few respondents were overlooking, not giving full attention and responded in haste without giving full attention to each and every questions. Such samples were not approved and disqualified. Table 4.1 shows the total number of samples contacted across the divisions, samples thus rejected and finally considered for analysis.

The responses were collected covering consumers of different age group. Table 4.2 shows the distribution of respondents as per age group.

Table 4.2: Age-wise distribution of the overall respondents in the study area

Age group	Frequency	Percentage
18-25	724	50.42
26-35	323	22.49
36-45	205	14.28
46-55	121	8.43
55-65	50	3.48
65 and Above	13	0.91
Total	1,436	100

Source: Survey data

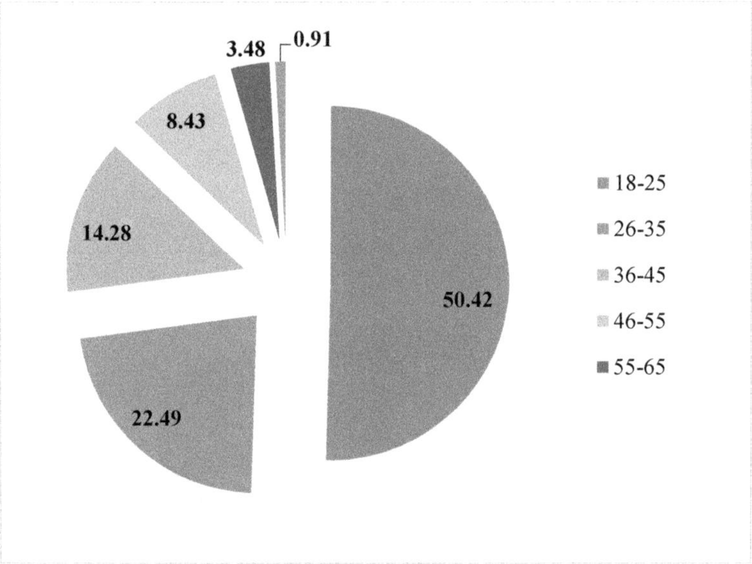

Fig. 4.2: Age-wise Distribution of the Respondents (%)

Source: Survey data

Table 4.3: Division-wise share of respondents as per age group

(figures in parentheses shows percentage)

Division	Age group						Total
	18-25	26-35	36-45	46-55	55-65	65 and Above	
Bengaluru	276 (48.25)	169 (29.65)	74 (12.99)	33 (5.79)	13 (2.29)	6 (1.06)	571 (100)
Mysuru	187 (66.32)	39 (13.83)	26 (9.22)	23 (8.16)	7 (2.49)	0 (0)	282 (100)
Belagavi	119 (43.92)	41 (15.13)	60 (22.15)	34 (12.55)	13 (4.8)	4 (1.48)	271 (100)
Gulbarga	143 (45.84)	74 (23.72)	45 (14.43)	31 (9.94)	16 (5.13)	3 (0.97)	312 (100)
Total	725 (50.46)	323 (22.51)	205 (14.29)	121 (8.44)	49 (3.42)	13 (0.91)	1436 (100)

Source: Survey data

From the Table 4.3, it has been found that the majority of the respondents (around 73 percent) are from the age group up to 35, other 26 percent respondents come under the age group of 36-45, 46-55 and 55-65, further a very few respondents i.e. around 1 per cent are from the age group above 65. Therefore we can say that the young generations are more enthusiastic to participate in the present study rather than the mid and old age group of the respondents.

If we look into the region-wise age distribution of the respondents, table 4.2 shows that almost the similar distribution has been followed. In the Bengaluru region, it has been found that around 78 percent of the respondents are from the age group up to 35, followed by around 19 percent respondents under the age group of 36-45 and 46-55, rest 3.35 percent respondents are from the age group 55-65 and above 65. In case of the Mysore region, it has been observed from the study that 80 percent of the respondents are from the age group up to 35, followed by around 17 percent respondents are come under the age group of 36-45 and 46-55, further rest 2.49 percent respondents are from the age group 55-65 and no respondents has been found in the age group of above 65. Again, in the Belgaum region 59 percent of respondents come under the age group up to 35, while around 35 percents are under the age group of 36-45 and 46-55 and the rest 6.28 percent of the respondents are from the age group of 55-65 and above 65. Lastly, it has been observed from the study that in case of Gulbarga region around 70 percent of the respondents are from the age group up to 35, followed by 24 percent respondents under the age group of 36-45 and 46-55, further rest 6 percent respondents are from the age group 55-65 and above 65.

Table 4.4: shows the distribution of respondents as per gender.

Table 4.4: Gender-wise share of the overall respondents in the study area

Gender	Freq.	Percent
Male	852	59.33
Female	584	40.67
Total	1,436	100

Source: Survey data

Table 4.5: Division-wise gender profile of the Respondents

(figures in parentheses shows percentage)

	Gender		
Division	**Male**	**Female**	**Total**
Bengaluru	340(59.55)	231(40.46)	571(100)
Mysuru	124(43.98)	158(56.03)	282(100)
Belagavi	156(57.57)	115(42.44)	271(100)
Gulbarga	232(74.36)	80(25.65)	312(100)
Total	852(59.34)	584(40.67)	1436(100)

Source: Survey data

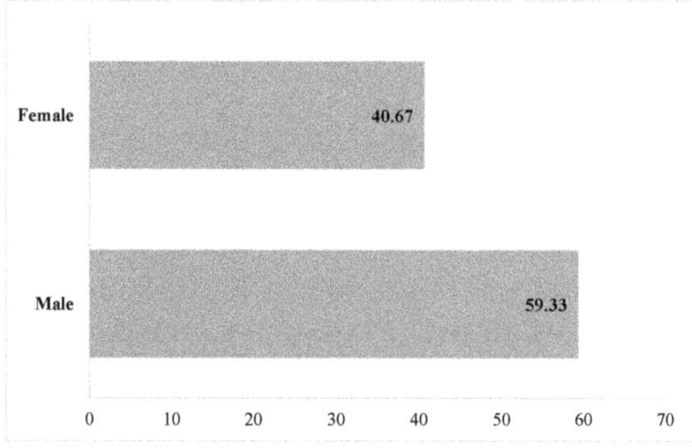

Fig. 4.3: Gender-wise Distribution of the Respondents (%)

Source: Survey data

It has been observed from the table 4.4 that, male respondents were more than the female respondents. Where, 59 per cent were the male respondents and around 41 per cent were the female respondents.

If we look into the division-wise gender profile of the respondents, it has been portrayed from the study that in Bengaluru division around 60 percent are male respondents and rest around 40 percent are female.. In Mysore division the opposite picture has been observed where female respondents (56 %) are bit higher than the male respondents (44%). Further, in the Belgaum division around 58 percent are male respondents and rest 42 percent are females. In case of Gulbarga division, it has been observed from the study that 74 percent are male respondents and around 26 percent are female respondents. Therefore, if we look into the overall result of the study in terms of gender participation male respondents has participated more actively than the female respondents.

The distribution of respondents as per their marital status is shown the table 4.5.

Table 4.5: Marital Status of the respondent

Marital Status	Frequency	Percent
Single	810	56.41
Married	625	43.52
Total	1,436	100.00

Source: Survey data

Table 4.6: Division-wise Marital Status of the Respondents

(figures in parentheses shows percentage)

Division	Marital Status		Total
Bengaluru	315(55.17)	256(44.84)	571(100)
Mysuru	194(68.8)	88(31.21)	282(100)
Belagavi	136(50.19)	135(49.82)	271(100)
Gulbarga	166(53.21)	146(46.8)	312(100)
Total	811(56.48)	625(43.53)	1436(100)

Source: Survey data

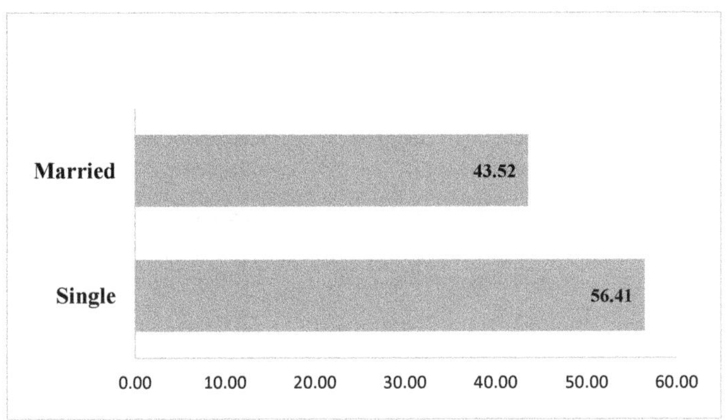

Fig. 4.4: Marital Status-wise Distribution of the Respondents (%)

Source: Survey data

It has been found table 4.6 that a majority of 56 percent of respondents are single and around 44 percent respondents are married.

If we look into the division-wise status, all the divisions a similar distribution pattern of respondents are found. Where, in Bengaluru division it has been found that 55 percent of the respondents are single and rest aroud 45 percent are married. In case of Mysore division, it has been observed that around 69 percent of the respondents are single and 31 percent has been found as married. Again, in case of the Balgaum, equal participaton of both category has been observed with 50 percent in each category. Lastly, it has been noticed in case of Gulbarga division that 53 percent of respondents come under single and rest around 47 percent are married respondents.

Table 4.7 shows the distribution of respondents as per educational qualification.

Table 4.7: Educational qualification of the respondents

Education	Respondents	
	Frequency	Percent
No formal Education	96	6.69
Up to High School	249	17.34
Bachelor's degree	552	38.44
Master's/ Professional Degree	460	32.03
Others	79	5.5
Total	1,436	100

Source: Survey data

Table 4.8: Division-wise Education qualification of the Respondents

(figures in parentheses shows percentage)

Division	Education Qualification					Total
	No formal Education	Up-to High School	Bachelor degree	Master/ Professional Degree	Others	
Bengaluru	28(4.91)	59(10.34)	240(42.04)	185(32.4)	59(10.34)	571(100)
Mysuru	4(1.42)	72(25.54)	49(17.38)	155(54.97)	2(0.71)	282(100)
Belagavi	35(12.92)	63(23.25)	120(44.29)	48(17.72)	5(1.85)	271(100)
Gulbarga	29(9.3)	55(17.63)	143(45.84)	72(23.08)	13(4.17)	312(100)
Total	96(6.69)	249(17.34)	552(38.45)	460(32.04)	79(5.51)	1436(100)

Source: Survey data

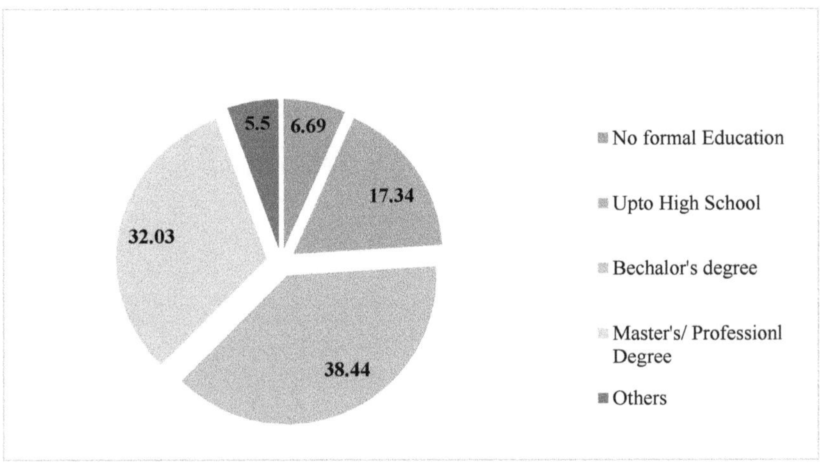

Fig. 4.5: Education-wise Distribution of the Respondents (%)

Source: Survey data

If we look into the educational qualification of the respondents, highest i.e. 38 per cent respondents hold bachelor's degree followed by Master's / professional degree holders with 32 percent, up to high school educated with 17 per cent, around 7 percent having no formal education and rest 5.5 percent respondents were holding other educational qualification.

If we look into the division-wise figure, in Bengaluru, highest i.e. 42 per cent respondents holding bachelor's degree followed by Master's / professional degree holders with 32 percent, 10 percent each respondents area educated up to high school and others, respectively. Lastly, around 5 percent respondents are not having any formal education. It has been observed in Mysore division that, highest number i.e. around 55 per cent respondents are holding Master's / professional degree followed by up to high school degree with around 26 percent, 17 percent respondents are bachelor's degree holders and 1.42 are come under the category of no formal education. Lastly, a very few respondents which is on around 1 percent are having other education qualification. Again, in Belgaum division highest number i.e. 44 per cent respondents are holding bachelor's degree, followed by educated up to high school with 23 percent, around 18 percent respondents has been found with holding Master's / professional degree and around 13 percent respondents are no formal educated, respectively. Lastly, around 2 percent respondents are having other education qualification. Further, if we look into the Gulbarga division highest number i.e. around 46 per cent of the respondents are holding bachelor's degree, followed by respondents are holding Master's / professional degree with around 23 percent, around 18 percent respondents has been found with holding up-to high school degree and around 9 percent respondents are having no formal education, respectively. Lastly, around 4 percent respondents are having other education qualification.

Table 4.9 shows the distribution of respondents as per their occupation.

Table 4.9: Occupation of the respondents

Occupation	Respondents	
	Frequency	Percent
Student	637	44.36
Home maker	181	12.6
Agri. And Allied Activities	121	8.43
Salaried entrepreneurs/ Self Business	398	27.72
Others	99	6.89
Total	1,436	100

Source: Survey data

Table 4.10: Division-wise Occupation details of the Respondents

(figures in parentheses shows percentage)

Division	Occupation					Total
	Student	Home maker	Agri. And Allied Activities	Salaried entrepreneurs/ Self Business	Others	
Bengaluru	241(42.21)	74(12.96)	62(10.86)	141(24.7)	53(9.29)	571(100)
Mysore	176(62.42)	14(4.97)	6(2.13)	69(24.47)	17(6.03)	282(100)
Belgaum	112(41.33)	54(19.93)	29(10.71)	70(25.84)	6(2.22)	271(100)
Gulbarga	108(34.62)	39(12.5)	24(7.7)	118(37.83)	23(7.38)	312(100)
Total	637(44.36)	181(12.61)	121(8.43)	398(27.72)	99(6.9)	1436(100)

Source: Survey data

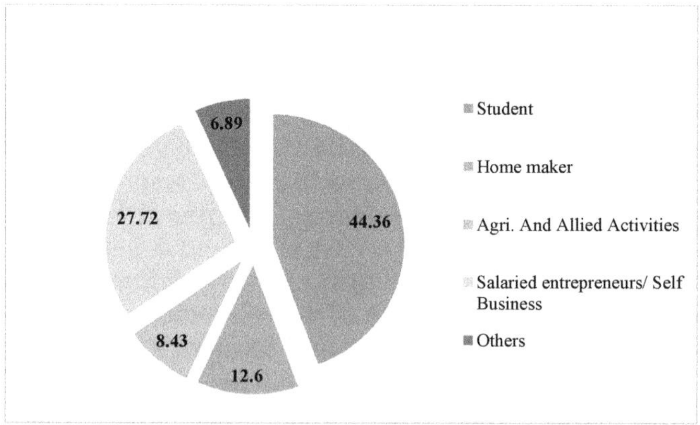

Fig. 4.6: Occupation-wise distribution of the Respondents (%)

Source: Survey data

From the table 4.9, it has been found that the majority of respondents i.e. 44 percent are students, followed by having the occupation of Salaried entrepreneurs/ Self Business salaried with around 28, around 13 percent are come under the category of home maker, 8 percent respondents are engaged in agri. and allied activities and rest around 7 come under the other occupation category.

In this section (table 4.10), it has been found that in Bengaluru division the majority of respondents i.e. 42 percent are students, followed by having the occupation of Salaried entrepreneurs/ Self Business salaried with around 25, around 13 percent are come under the category of home maker, around 11 percent respondents are engaged in agri. and allied activities and rest 9 percent respondents of the study come under the other occupation category. Again, in Mysore division the majority of respondents i.e. 62 percent are students, followed by having the occupation of Salaried entrepreneurs/ Self Business salaried with around 25, 6 percent are come under the category of other occupation, around 5 percent respondents are home maker and rest 2 percent respondents of the study engaged in agri. and allied activities. Where, in case of Belgaum, majority of respondents i.e. 41 percent are students, followed by having the occupation of Salaried entrepreneurs/ Self Business salaried with around 26, around 20 percent respondents are home maker, where around 11 percent respondents are engaged in agri. and allied activities and other 2 percent respondents of the study are come under the category of other occupation. Lastly, in Gulbarga majority of respondents i.e. around 38 percent are having the occupation of Salaried entrepreneurs/ Self Business salaried, followed by around 35 percent respondents are students, where around 13 percent respondents are come under the category of home maker, and around 8 and 7 percent each respondents are engaged in agri. and allied activities and other occupation category, respectively.

From the overall observation regarding occupation of the respondents, majority of them are students and salaried professionals and least number of respondents.

The table 4.11 shows the distribution of respondents as per their annual family income.

Table 4.11: Annual family income of the respondents

Annual Family Income	**Respondents**	
	Frequency	**Percent**
Less than Rs. 1,00,000	662	46.1
Rs. 1,00,000 to less than Rs. 4,00,000	476	33.15
Rs. 4,00,000 to less than Rs. 7,00,000	178	12.4
Rs. 7,00,000 to less than Rs. 10,00,000	58	4.04
Above Rs. 10,00,000	62	4.32
Total	1,436	100

Source: Survey data

Table 4.12: Division-wise Annual family Income details of the Respondents

(figures in parentheses shows percentage)

Division	Annual Family Income					Total
	> Rs. 1,00,000	Rs. 1,00,000 to > Rs. 4,00,000	Rs. 4,00,000 to > Rs. 7,00,000	Rs. 7,00,000 to > Rs. 10,00,000	<Rs. 10,00,000	
Bengaluru	271(47.47)	164(28.73)	63(11.04)	32(5.61)	41(7.19)	571(100)
Mysore	115(40.79)	76(26.96)	61(21.64)	16(5.68)	14(4.97)	282(100)
Belgaum	117(43.18)	114(42.07)	36(13.29)	1(0.37)	3(1.11)	271(100)
Gulbarga	159(50.97)	122(39.11)	18(5.77)	9(2.89)	4(1.29)	312(100)
Total	662(46.11)	476(33.15)	178(12.4)	58(4.04)	62(4.32)	1436(100)

Source: Survey data

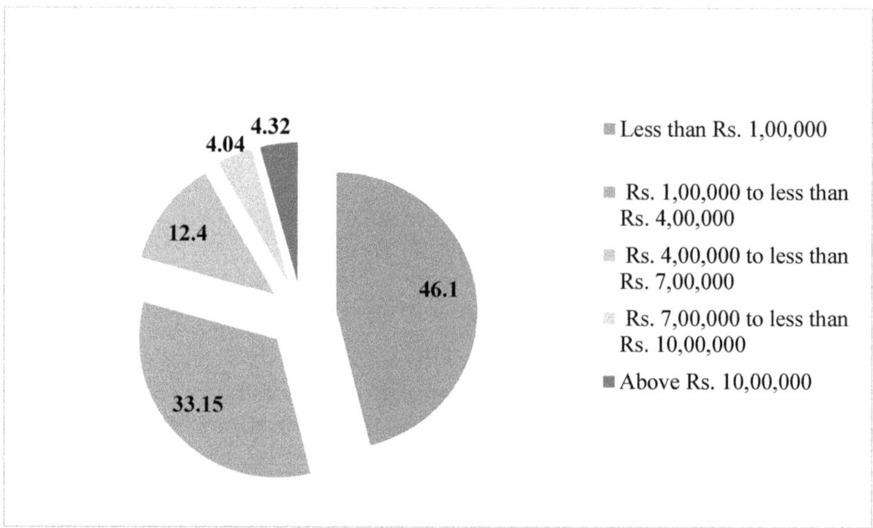

Fig. 4.7: Family Income-wise Distribution of the Respondents (%)

Source: Survey data

From the table 4.11, it has been found the majority of the respondents i.e.46 per cent come under the annual income category of less than Rs. 1,00, 000, followed by 33 per cent respondents under the annual income category of Rs. 1,00,000 to less than Rs. 4,00,000, 12 per cent respondents come under the annual income category of Rs. 4,00,000 to less than Rs. 7,00,000 and further 4 per cent each respondents are come under the annual income group category of Rs. 7,00,000 to less than Rs. 10,00,000 and Above Rs. 10,00,000.

The division-wise picture has been depicted in the table 4.12, where in Bengaluru the majority of the respondents i.e. 47 per cent respondents come under the annual income category of less than Rs. 1, 00, 000, followed by around 29 per cent respondents under the annual income category of Rs. 1,00,000 to less than Rs. 4,00,000, 11 per cent respondents come under the annual income category of Rs. 4,00,000 to less than Rs. 7,00,000, 7 per cent respondents are come under the annual income group category of Above Rs. 10,00,000 and lastly around 6 percent respondents are come under the income group category of Rs. 7,00,000 to less than Rs. 10,00,000. In case of Mysore, the majority of the respondents i.e. 41 per cent respondents come under the annual income category of less than Rs. 1, 00, 000, followed by around 27 per cent respondents under the annual income category of Rs. 1,00,000 to less than Rs. 4,00,000, 22 per cent respondents come under the annual income category of Rs. 4,00,000 to less than Rs. 7,00,000, around 6 per cent respondents are come under the annual income group category of Rs.7,00,000 to less than Rs.10,00,000 and lastly around 5 percent respondents are come under the income group category of Above Rs.10,00,000. Further, in Belgaum the majority of the respondents i.e. 43 per cent respondents are come under the annual income category of less than Rs. 1,00, 000, followed by around 42 per cent respondents under the annual income category of Rs.1,00,000 to less than Rs.4,00,000, 13 per cent respondents come under the annual income category of Rs. 4,00,000 to less than Rs. 7,00,000, 1 per cent respondents are come under the annual income group category of Above Rs. 10,00,000 and lastly a very few respondents i.e. 0.37 percent are come under the income group category of Rs. 7,00,000 to less than Rs. 10,00,000. Again, in Gulbarga the majority of the respondents i.e. around 51 per cent respondents come under the annual income category of less than Rs. 1, 00, 000, followed by around 39 per cent respondents under the annual income category of Rs. 1,00,000 to less than Rs. 4,00,000, around 6 per cent respondents come under the annual income category of Rs. 4,00,000 to less than Rs. 7,00,000, around 3 per cent respondents come under the annual income group category of Rs. 7,00,000 to less than Rs. 10,00,000 and lastly 1 percent respondents come under the income group category of Above Rs. 10,00,000.

In this section we can make a conclusion that majority of the respondents come under the annual income group of less than Rs. 1, 00, 000, this might be due to the fact that majority of the respondents' occupation is student.

Table 4.13 shows the distribution of respondents as per their place of residence.

Table 4.13: Place of residence of the respondents

Place of Residence	Respondents	
	Frequency	Percent
Rural	611	42.55
Urban	825	57.45
Total	1,436	100

Source: Survey data

Table 4.14: Division-wise respondents' place of residence

(figures in parentheses shows percentage)

Division	Place of Residence		Total
	Rural	Urban	
Bengaluru	237(41.51)	334(58.5)	571(100)
Mysore	167(59.22)	115(40.79)	282(100)
Belgaum	104(38.38)	167(61.63)	271(100)
Gulbarga	103(33.02)	209(66.99)	312(100)
Total	611(42.55)	825(57.46)	1436(100)

Source: Survey data

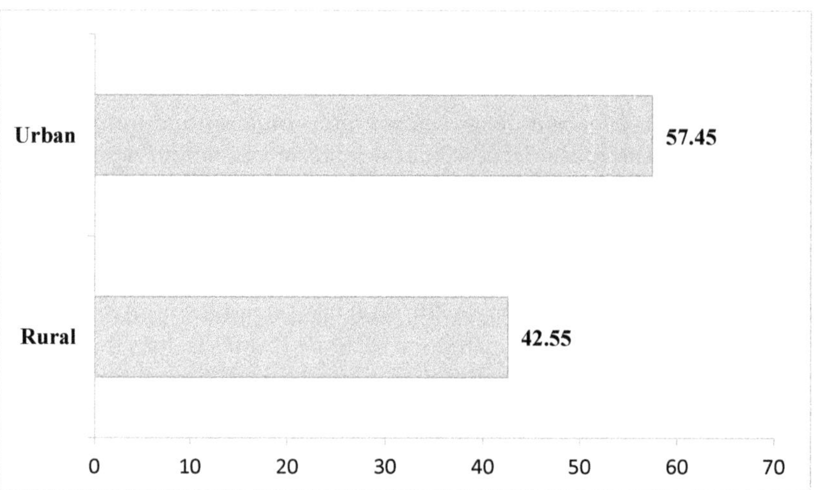

Fig. 4.8: Residential Area-wise Distribution of the Respondents (%)

Source: Survey data

From the table 4.13 it has been found that majority (57 per cent) of the respondents are from urban area and rest 43 per cent respondents belong to the rural area.

In Bengaluru division, around 59 percent of the respondents residing in urban area and rest aroud 42 percent in rural. In case of Mysore division, 59 percent of the respondents are belongs to the urban residential category and other around 41 percent from the rural residential background. Again, in case of Belgaum, the result is quite opposite then earlier two region, where the rural background respondents (62%) are more than the urban (38%) residents. Lastly, in Gulbarga division, 67 percent respondent are from rural background and rest 33 percent are from urban residential background.

4.2. Consumer Awareness level on Coffee/Tea

The consumers' awareness level on various aspects of the products play a significant role in their further evaluation of the product, brand preferences and an intention to buy. Consumers tend to buy a familiar and well known product. Brand awareness can help consumers to recognize a brand from a product category and make purchase decision. An understanding on this awareness level will help the policy makers to initiate required policy decisions and marketers to make appropriate marketing decisions. Therefore, the respondents were asked to mark their awareness level on the places where coffee/tea is grown globally and the available brands and quality.

Awareness on Places in India where Coffee and Tea Grown

Fig. 4: shows the awareness level of consumers on the places in Indian where coffee and tea is grown.

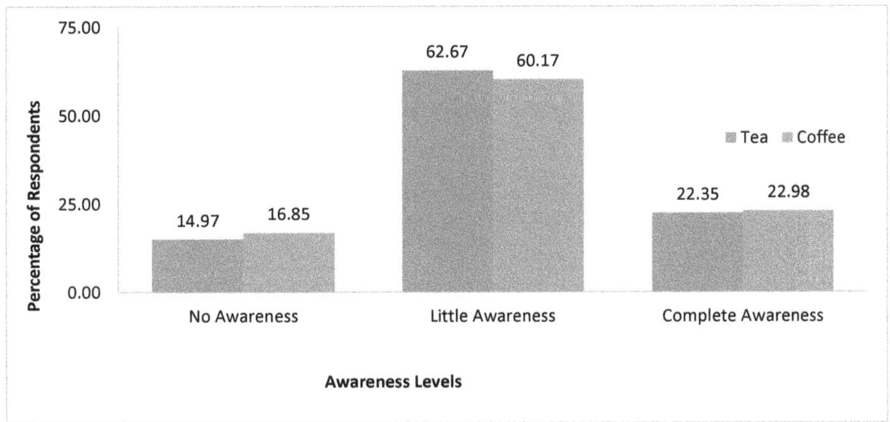

Fig. 4.9: Awareness Levels on the places in India where Tea and Coffee grown

The study revealed that the consumers' awareness level does not vary much for places in India where coffee and tea grown. A t-test paired comparison for the mean of coffee and tea has been done and it revealed that, invariably for all demography classification, the calculated value of p ◎ 0.05, therefore, there is no significant difference in the awareness level on coffee producing area and tea producing area in India. Hardly, about 22 % of respondents have the complete awareness on the places in India where coffee and tea is grown. Rests of the majority either have little or no awareness. This shows that the consumers were consuming tea and coffee without knowing much about the places in India where they are grown.

To find out, whether the awareness levels vary according to the demography, the data are presented accordingly and analyzed. Table 4.15.shows the awareness level of consumers on the places in Indian where coffee and tea is grown as per the demography.

Table 4.15: Consumers' Awareness Level on Places in India where coffee and tea is grown

	Number of Respondents						
	No awareness		Little awareness		Complete awareness		Total
	Tea	Coffee	Tea	Coffee	Tea	Coffee	Tea/ Coffee
Age Group							
18-25	61 28.37%	93 (12.85)	498 (68.78)	466 (64.36)	165 (22.79)	165 (22.79)	724 (100)
26-35	58 (17.96)	61 (18.89)	190 (58.82)	185 (57.28)	75 (23.22)	77 (23.84)	323 (100)
36-45	57 (27.80)	51 (24.87)	114 (55.60)	115 (56.09)	34 (16.58)	39 (19.02)	205 (100)
46-55	29 (23.96)	25 (20.66)	62 (51.23)	64 (52.89)	30 (24.79)	32 (26.44)	121 (100)
55-65	6 (12.00)	8 (16.00)	32 (64.00)	30 (60.00)	12 (24.00)	12 (24.00)	50 (100)
65 and Above	4 (30.76)	4 (30.76)	4 (30.76)	4 (30.76)	5 (38.46)	5 (38.46)	13 (100)
Total	215 (14.97)	242 (16.85)	900 (62.67)	864 (60.17)	321 (22.35)	330 (22.98)	1436 (100)
Gender							
Male	133 (15.61)	150 (17.60)	517 (60.68)	496 (58.21)	202 (23.70)	206 (24.17)	852 (100)
Female	82 (14.04)	92 (15.75)	383 (65.58)	368 (63.01)	119 (20.37)	124 (21.23)	584 (100)
Total	215(14.97)	242 (16.85)	900 (62.67)	864 (60.17)	321 (22.35)	330 (22.98)	1436 (100)
Marital Status							
Single	73 (9.01)	109 (13.45)	549 (67.77)	519 (64.07)	188 (23.20)	182 (22.46)	810 (100)
Married	142 (22.68)	133 (21.24)	351 (56.07)	345 (55.11)	133 (21.24)	148 (23.64)	626 (100)
Total	215 (14.97)	242 (16.85)	900 (62.67)	864 (60.17)	321 (22.35)	330 (22.98)	1436 (100)
Education							
No formal Education	35 (36.45)	36 (37.50)	52 (54.16)	48 (50.00)	9 (9.37)	12 (12.50)	96 (100)

Upto High School	67 (26.90)	76 (30.52)	145 (58.23)	130 (52.20)	37 (14.85)	43 (17.26)	249 (100)
Bachelor degree	71 (12.86)	76 (13.76)	362 (65.57)	353 (63.94)	119 (21.55)	123 (22.28)	552 (100)
Master/ Professional Degree	30 (6.52)	44 (9.56)	294 (63.91)	281 (61.08)	136 (29.56)	135 (29.34)	460 (100)
Others	12 (15.19)	10 (12.66)	47 (59.49)	52 (65.82)	20 (25.32)	17 (21.52)	79 (100)
Total	215 (14.97)	242 (16.85)	900 (62.67)	864 (60.17)	321 (22.35)	330 (22.98)	1436 (100)
Occupation							
Student	43 (6.75)	68 (10.67)	444 (69.70)	426 (66.87)	150 (23.54)	143 (22.44)	637 (100)
Home maker	56 (30.93)	51 (28.17)	105 (58.01)	102 (56.53)	20 (11.04)	28 (15.46)	181 (100)
Agri. And Allied Activities	37 (30.57)	40 (33.05)	59 (48.76)	52 (42.97)	25 (20.66)	29 (22.48)	121 (100)
Salaried entrepreneurs/ Self Business	76 (19.09)	76 (19.09)	231 (58.04)	227 (57.03)	91 (22.86)	95 (23.86)	398 (100)
Others	3 (3.03)	7 (7.07)	61 (61.61)	57 (57.58)	35 (35.35)	35 (35.35)	99 (100)
Total	215 (14.97)	242 (16.85)	900 (62.67)	864 (60.17)	321 (22.35)	330 (22.98)	1436 (100)
Annual Family Income							
Less than Rs. 1,00,000	111 (16.76)	127 (19.18)	425 (64.19)	402 (60.72)	126 (19.03)	133 (20.09)	662 (100)
Rs. 1,00,000 to less than Rs. 4,00,000	82 (17.22)	81 (17.01)	302 (63.44)	307 (64.49)	92 (19.32)	88 (18.48)	476 (100)
Rs. 4,00,000 to less than Rs. 7,00,000	20 (11.23)	24 (13.48)	101 (56.74)	100 (56.17)	57 (32.02)	54 (30.33)	178 (100)
Rs. 7,00,000 to less than Rs. 10,00,000	1 (1.72)	5 (8.62)	37 (63.79)	26 (44.82)	20 (34.48)	27 (46.55)	58 (100)
Above Rs. 10,00,000	1 (1.61)	5 (8.06)	35 (56.45)	29 (46.77)	26 (41.93)	28 (45.16)	62 (100)
Total	215 (14.97)	242 (16.85)	900 (62.67)	864 (60.17)	321 (22.35)	330 (22.98)	1436 (100)

Residence Location							
Rural	85 (13.91)	110 (18.00)	382 (62.52)	363 (59.41)	144 (23.56)	138 (22.58)	611 (100)
Urban	130 (15.75)	132 (16.00)	518 (62.78)	501 (60.72)	177 (21.45)	192 (23.27)	825 (100)
Total	215 (14.97)	242 (16.85)	900 (62.67)	864 (60.17)	321 (22.35)	330 (22.98)	1436 (100)
Division							
Bengaluru	73 (12.78)	70 (12.25)	359 (62.87)	348 (60.95)	139 (24.34)	153 (26.80)	571 (100)
Mysuru	9 (3.20)	23 (8.16)	167 (59.22)	159 (56.38)	106 (37.59)	100 (35.46)	282 (100)
Belagavi	82 (30.26)	88 (32.47)	160 (59.04)	158 (58.30)	29 (10.70)	25 (9.23)	271 (100)
Gubarga	51 (16.35)	61 (19.55)	214 (68.59)	199 (63.78)	47 (15.06)	52 (16.67)	312 (100)
Total	215 (14.97)	242 (16.85)	900 (62.67)	864 (60.17)	321 (22.35)	330 (22.98)	1436 (100)

To find out whether there is any significant variation in the awareness level across the demography, analysis of variance is done on the calculated mean separately for coffee and tea.

Table 4.16: ANOVA Single factor for variation in awareness level on places in India coffee/tea grown across demography

Demography	Calculated p value		Result (significant if $p < 0.05$ at 95%)
	Tea	Coffee	
Age group	8.38	1.66	Insignificant
Gender	0.000569	0.000673	Significant
Marital Status	0.036076	0.005929	
Education	9.02	2.32	Insignificant
Occupation	4.97	6.09	
Annual family income	7.27	7.45	
Place of residence	5.79	5.34	
Division	0.000159	0.001267	Significant

Though the overall percentage of respondents who have a complete awareness on the places where coffee and tea grown is found to be very less, the awareness level vary as per the age group. The respondents above the age of 45 have marginally more awareness on the same. On the other hand, the awareness level does not vary much as per the gender and marital status. The uneducated respondents relatively have a poor awareness level, hardly around 9.37% of respondents have complete awareness on the places in India where tea is grown, likewise, 12.50% of respondents for coffee. The awareness level increases with the educational advancement, with about 30% of the respondents having complete awareness on the places where coffee and tea is grown in India.

It is quite interesting to infer from the table that the percentage of respondents who are having a complete awareness level on the places where coffee and tea is grown in India does not vary much as per the occupation, except the 'home maker' category. The home makers relatively have a poor awareness (15.46%) on the same. Incidentally, for products like coffee and tea, home makers are the one who makes the shopping decision, unfortunately having poor awareness. There is a strong pattern of awareness level as per the annual family income. Family having annual income less than Rs.1,00,000, have a poor awareness level (@ 20%) and the level of awareness increases with income leading to about 45% of the respondents having a complete awareness. Though there is an insignificant variation in the awareness level as per the place of residence of respondents, the urbanites have slightly better awareness than their rural counterparts.

Awareness on Countries where Coffee and Tea Grown

As we are in the boundary less globalized market, coffee and tea produced across the globe are expected to be marketed in the domestic market. Coffee and tea produced across different countries vary in their taste, flavor, aroma, etc. and most importantly in their safety, as they are produced in varied agro-climatical conditions and processed under varying regulatory standards across different countries. Therefore, Indian consumers are expected to be aware of the countries where coffee and tea is produced.

The respondents were asked to mark their awareness level on the various countries where coffee and tea is produced. The figure 4.10.shows the details.

It is made evident from the study that the awareness level on the countries where coffee and tea is produced is very less. And the awareness level does not vary much for coffee and tea. A t-test paired comparison for the mean of coffee and tea has been done and it revealed that, invariably for all demography classification, the calculated value of p ◎ 0.05, therefore, there is no significant difference in the awareness level on countries where tea is produced and coffee is produced. The percentage of respondents having complete awareness for coffee is 12.81 % which is slightly higher than that of tea (10.86%). This may be because, the study area, coffee consumers are more than the tea consumers. About 36% of them do not have any idea on the countries where coffee and tea is grown. To find out, whether the awareness level vary according to the demography, the data are presented accordingly and analyzed.

Table 4.17: Consumers' Awareness Level on countries where coffee and tea is grown

	Number of Respondents						
	No awareness		Little awareness		Complete awareness		Total
	Tea	Coffee	Tea	Coffee	Tea	Coffee	Tea/Coffee
Age group							
18-25	209 (28.87)	222 (30.66)	431 (59.53)	398 (54.97)	84 (11.60)	104 (14.36)	724 (100)
26-35	121 (37.46)	129 (39.94)	167 (51.70)	167 (51.70)	35 (10.84)	27 (8.36)	323 (100)
36-45	86 (41.95)	82 (40.00)	105 (51.22)	101 (49.27)	14 (6.83)	22 (10.73)	205 (100)
46-55	48 (39.67)	55 (45.45)	57 (47.11)	44 (36.36)	16 (13.22)	22 (18.18)	121 (100)
55-65	20 (40.00)	20 (40.00)	24 (48.00)	22 (44.00)	6 (12.00)	8 (16.00)	50 (100)
65 and Above	5 (38.46)	7 (53.85)	7 (53.85)	5 (38.46)	1 (7.69)	1 (7.69)	13 (100)
Total	489 (34.05)	515 (35.86)	791 (55.08)	737 (51.32)	156 (10.86)	184 (12.81)	1436 (100)
Gender							
Male	296 (34.74)	318 (37.32)	457 (53.64)	417 (48.94)	99 (11.62)	117 (13.73)	852 (100)
Female	193 (33.05)	197 (33.73)	334 (57.19)	320 (54.79)	57 (9.76)	67 (11.47)	584 (100)
Total	489 (34.05)	515 (35.86)	791 (55.08)	737 (51.32)	156 (10.86)	184 (12.81)	1436 (100)
Marital Status							
Single	242 (29.88)	259 (31.98)	476 (58.77)	443 (54.69)	92 (11.36)	108 (13.33)	810 (100)
Married	247 (39.46)	256 (40.89)	315 (50.32)	294 (46.96)	64 (10.22)	76 (12.14)	626 (100)
Total	489 (34.05)	515 (35.86)	791 (55.08)	737 (51.32)	156 (10.86)	184 (12.81)	1436 (100)
Education							
No formal Education	51 (53.13)	54 (56.25)	37 (38.54)	37 (38.54)	8 (8.33)	5 (5.21)	96 (100)
Upto High School	114 (45.78)	120 (48.19)	111 (44.58)	100 (40.16)	24 (9.64)	29 (11.65)	249 (100)
Bachelor degree	198 (35.87)	210 (38.04)	302 (54.71)	274 (49.64)	52 (9.42)	68 (12.32)	552 (100)

Master/ Professional Degree	96 (20.87)	97 (21.09)	302 (65.65)	289 (62.83)	62 (13.48)	74 (16.09)	460 (100)
Others	30 (37.97)	34 (43.04)	39 (49.37)	37 (46.84)	10 (12.66)	8 (10.13)	79 (100)
Total	489 (34.05)	515 (35.86)	791 (55.08)	737 (51.32)	156 (10.86)	184 (12.81)	1436 (100)
Occupation							
Student	172 (27.00)	180 (28.26)	385 (60.44)	364 (57.14)	80 (12.56)	93 (14.60)	637 (100)
Home maker	85 (46.96)	89 (49.17)	86 (47.51)	82 (45.30)	10 (5.52)	10 (5.52)	181 (100)
Agri. And Allied Activities	62 (51.23)	65 (53.71)	45 (37.19)	43 (35.53)	14 (11.57)	13 (10.74)	121 (100)
Salaried entrepreneurs/ Self Business	136 (34.17)	145 (36.43)	218 (54.77)	199 (50.00)	44 (11.05)	54 (13.56)	398 (100)
Others	34 (34.34)	36 (36.36)	57 (57.57)	49 (49.49)	8 (8.08)	14 (14.14)	99 (100)
Total	489 (34.05)	515 (35.86)	791 (55.08)	737 (51.32)	156 (10.86)	184 (12.81)	1436 (100)
Annual Family Income							
Less than Rs. 1,00,000	262 (39.57)	285 (43.05)	335 (50.60)	303 (45.77)	65 (9.81)	74 (11.17)	662 (100)
Rs. 1,00,000 to less than Rs. 4,00,000	162 (34.03)	170 (35.71)	272 (57.14)	262 (55.04)	42 (8.82)	44 (9.24)	476 (100)
Rs. 4,00,000 to less than Rs. 7,00,000	44 (24.71)	43 (24.15)	105 (58.98)	102 (57.30)	29 (16.29)	33 (18.53)	178 (100)
Rs. 7,00,000 to less than Rs. 10,00,000	8 (13.79)	8 (13.79)	40 (68.96)	30 (51.72)	10 (17.24)	20 (34.48)	58 (100)
Above Rs. 10,00,000	13 (20.96)	9 (14.51)	39 (62.90)	40 (64.51)	10 (16.12)	13 (20.96)	62 (100)
Total	489 (34.05)	515 (35.86)	791 (55.08)	737 (51.32)	156 (10.86)	184 (12.81)	1436 (100)
Residence Location							
Rural	203 (33.22)	216 (35.35)	340 (55.64)	303 (49.59)	68 (11.12)	92 (15.05)	611 (100)
Urban	286 (34.66)	299 (36.24)	451 (54.66)	434 (52.60)	88 (10.66)	92 (11.15)	825 (100)
Total	489 (34.05)	515 (35.86)	791 (55.08)	737 (51.32)	156 (10.86)	184 (12.81)	1436 (100)

Division							
Bengaluru	203 (35.55)	192 (33.62)	303 (53.06)	299 (52.36)	65 (11.38)	80 (14.01)	571 (100)
Mysuru	44 (15.60)	51 (18.08)	187 (66.31)	175 (62.05)	51 (18.08)	56 (19.85)	282 (100)
Belagavi	136 (50.18)	136 (50.18)	116 (42.80)	117 (43.17)	19 (7.01)	18 (6.64)	271 (100)
Gubarga	106 (33.17)	136 (43.58)	185 (59.29)	146 (46.79)	21 (6.73)	30 (9.61)	312 (100)
Total	489 (34.05)	515 (35.86)	791 (55.08)	737 (51.32)	156 (10.86)	184 (12.81)	1436 (100)

To find out whether there is any significant variation in the awareness level on the producing countries of tea and coffee across the demography, analysis of variance is carried out on the calculated mean separately for coffee and tea.

Table 4.18: ANOVA Single factor for variation in awareness level on countries where coffee/tea grown across demography

NOVA Single factor for variation in awareness level on countries w	Calculated p value		Result (significant if $p < 0.05$ at 95%)
	Tea	Coffee	
Age group	5.36	4.81	Insignificant
Gender	0.000323	0.00212	Significant for both coffee and tea
Marital Status	0.008325	0.009654	
Education	0.000446	0.000755	
Occupation	6.75	2.41	Insignificant
Annual family income	1.46	0.000299	Significant for coffee and insignificant for tea
Place of residence	2.38	0.000784	
Division	0.000696	0.00585	Significant for both coffee and tea

The percentage of respondents who have a complete awareness on the countries where coffee and tea is grown is less than who are aware of the places in India. It is also noted from the data that there is no strong association between the awareness level on the countries where coffee and tea is grown and the age group. The percentages of male respondents (11.62%) have better awareness than female respondents (9.76%) on the countries where coffee and tea grown. There is a little association between the awareness level and the marital status. The awareness level on countries also varies as per the educational qualification, with 'no formal education' category around 6% and 'master degree/professional degree' around 16% having complete awareness. Students relatively have a better awareness than respondents belonging to other occupation. There found to be a strong association between the awareness level and the annual family income, with higher family income, the awareness level is also higher.

Awareness on Foreign Brands of Coffee and Tea

For a consumer to narrow down and choose a brand, the concerned brand should be in the awareness set. The respondents were asked to choose their awareness level on the available foreign brands of coffee and tea. The figure below shows their response detail.

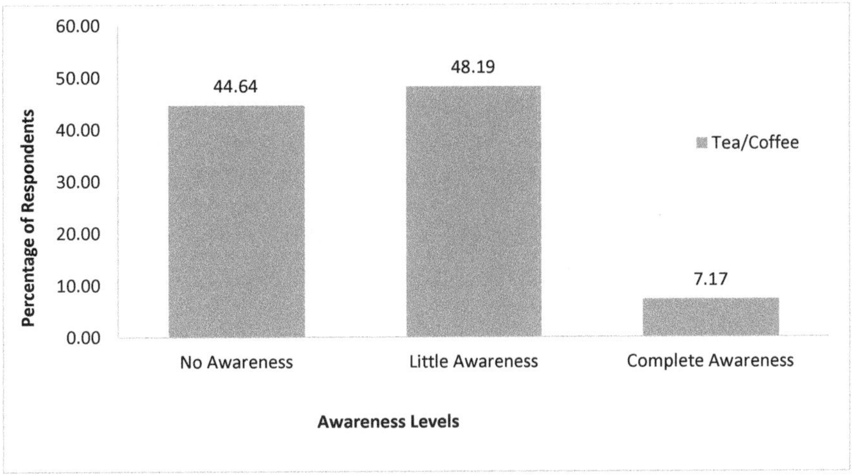

Fig. 4.11: Awareness Levels on Foreign brands of Coffee and Tea

The awareness level on the available foreign brands is very poor with just 7.17% of the respondents with complete awareness, 48.19% with little awareness and the rest 44.64% having no awareness.

To have a better understanding, analysis is made on the basis of demography and the same is presented below.

Table 4.19: Consumers' Awareness Level on the Foreign Brands of Coffee and Tea

	Number of Respondents			
	No awareness	**Little awareness**	**Complete awareness**	**Total**
Age Group				
18-25	292 (40.33)	366 (50.55)	66 (9.12)	724 (100)
26-35	153 (47.37)	153 (47.37)	17 (5.26)	323 (100)
36-45	111 (54.15)	83 (40.49)	11 (5.36)	205 (100)
46-55	52 (42.98)	61 (51.41)	8 (6.61)	121 (100)

55-65	27 (54.00)	22 (44.00)	1 (2.00)	50 (100)
65 and Above	6 (46.15)	7 (53.85)	0	13 (100)
Total	641 (44.64)	692 (48.19)	103 (7.17)	1436 (100)
Gender				
Male	371 (43.55)	421 (49.41)	60 (7.04)	852 (100)
Female	270 (46.23)	271 (46.40)	43 (7.36)	584 (100)
Total	641 (44.64)	692 (48.19)	103 (7.17)	1436 (100)
Marital Status				
Single	337 (41.61)	403 (49.75)	70 (8.64)	810 (100)
Married	304 (48.56)	289 (46.17)	33 (5.27)	626 (100)
Total	641 (44.64)	692 (48.19)	103 (7.17)	1436 (100)
Education				
No formal Education	65 (67.71)	28 (29.17)	3 (3.12)	96 (100)
Upto High School	124 (49.80)	104 (41.77)	21 (8.43)	249 (100)
Bachelor degree	247 (44.75)	272 (49.28)	33 (5.97)	552 (100)
Master/Professional Degree	155 (33.70)	260 (56.52)	45 (9.78)	460 (100)
Others	50 (63.30)	28 (35.44)	1 (1.26)	79 (100)
Total	641 (44.64)	692 (48.19)	103 (7.17)	1436 (100)
Occupation				
Student	259 (40.66)	320 (50.24)	58 (9.10)	637 (100)
Home maker	109 (60.22)	68 (37.57)	4 (2.21)	181 (100)
Agri. And Allied Activities	77 (63.64)	39 (32.23)	5 (4.13)	121 (100)

Salaried entrepreneurs/ Self Business	155 (38.94)	216 (54.27)	27 (6.78)	398 (100)
Others	41 (41.40)	49 (49.50)	9 (9.10)	99 (100)
Total	641 (44.64)	692 (48.19)	103 (7.17)	1436 (100)
Annual Family Income				
Less than Rs. 1,00,000	326 (49.24)	296 (44.71)	40 (6.04)	662 (100)
Rs. 1,00,000 to less than Rs. 4,00,000	218 (45.80)	232 (48.74)	26 (5.46)	476 (100)
Rs. 4,00,000 to less than Rs. 7,00,000	61 (34.27)	97 (54.50)	20 (11.24)	178 (100)
Rs. 7,00,000 to less than Rs. 10,00,000	21 (36.21)	30 (51.72)	7 (12.07)	58 (100)
Above Rs. 10,00,000	15 (24.20)	37 (59.68)	10 (16.13)	62 (100)
Total	641 (44.64)	692 (48.19)	103 (7.17)	1436 (100)
Residence Location				
Rural	282 (46.15)	283 (46.32)	46 (7.53)	611 (100)
Urban	359 (43.52)	409 (49.58)	57 (6.91)	825 (100)
Total	641 (44.64)	692 (48.19)	103 (7.17)	1436 (100)
Division				
Bengaluru	289 (50.61)	255 (44.66)	27 (4.73)	571 (100)
Mysuru	76 (26.95)	162 (57.45)	44 (15.60)	282 (100)
Belagavi	156 (57.46)	102 (37.64)	13 (4.80)	271 (100)
Gubarga	120 (38.46)	173 (55.45)	19 (6.09)	312 (100)
Total	641 (44.64)	692 (48.19)	103 (7.17)	1436 (100)

To find out whether there is any significant variation in the awareness level on the foreign brands of tea and coffee across the demography, analysis of variance is carried out on the calculated mean.

Table 4.20: ANOVA Single factor for variation in awareness level on foreign brands of coffee/tea across demography

Demography	*Calculated p value*	Result (significant if $p < 0.05$ at 95%)
Age group	1.95513	Insignificant
Gender	0.000248816	Significant
Marital Status	0.002216043	
Education	3.18942	
Occupation	1.0256	Insignificant
Annual family income	2.71879	
Place of residence	0.000286099	
Division	0.000478005	

The awareness level on the foreign brands of coffee and tea varies slightly as per the age group. The lower age group has relatively better awareness than higher age group respondents. The brand awareness also does not vary as per the gender. There found to be a significant difference in the awareness level as per the gender and marital status. Out of the 626 respondents who are under 'married' category only 33 (5.27%) respondents are having a complete awareness on the foreign brands. On the other hand, the percentage of respondents under 'single' category has relatively better awareness. Students, professionals and salaries class have better awareness than home makers and agriculture labourers. A maximum of 16.13% of respondents under the annual family income above Rs.10 lakhs have complete awareness, otherwise, the level of awareness level is very poor across other categories.

Awareness on Quality of Indian Coffee and Tea

Since coffee and tea are marketed as packed powder which is meant for consumption, the consumer is expected to be aware of the quality of the item. The percentage of respondents and their awareness level on the quality of coffee and tea produced and marketed in India is depicted in the following figure.

The figure shows that the Indian consumer's awareness level on the quality of coffee and tea produced in the country is very low. As many as 34% do not have any awareness at all and just about 19% have complete awareness. The coffee and tea produced in India has unique flavor, aroma and quality, making them quite unique. However, the low awareness level of consumers acts as a major barrier for promoting on country of origin labelling.

The awareness level on quality of tea produced in India and quality of coffee produced in India, a t-test of paired comparison on the average score of both has been done and it revealed that across the demography, the level of awareness is not varying significantly.

To find out, whether the awareness level on the quality of Indian coffee and tea vary according to the demography, data are presented accordingly and analyzed.

Table 4.21: Consumers' Awareness Level on the Quality of Indian Coffee and Tea

	Number of Respondents						
	No awareness		**Little awareness**		**Complete awareness**		**Total**
	Tea	**Coffee**	**Tea**	**Coffee**	**Tea**	**Coffee**	**Tea**
Age Group							
18-25	190 (26.24)	194 (26.80)	342 (47.24)	377 (52.07)	192 (26.52)	153 (21.13)	724 (100)
26-35	130 (40.25)	128 (39.63)	145 (44.89)	145 (44.890	48 (14.86)	50 (15.48)	323 (100)
36-45	93 (45.37)	95 (46.34)	92 (44.88)	87 (42.44)	20 (9.76)	23 (11.22)	205 (100)
46-55	54 (44.63)	55 (45.46)	54 (44.63)	43 (35.54)	13 (10.73)	23 (19.00)	121 (100)
55-65	19 (38.00)	17 (34.00)	27 (54.00)	22 (44.00)	4 (8.00)	11 (22.00)	50 (100)
65 and Above	4 (30.77)	7 (53.85)	4 (30.77)	2 (15.38)	5 (38.46)	4 (30.77)	13 (100)
Total	490 (34.12)	496 (34.54)	664 (46.24)	676 (47.08)	282 (19.64)	264 (18.38)	1436 (100)
Gender							
Male	294 (34.51)	309 (36.27)	388 (45.54)	389 (45.66)	170 (19.95)	154 (18.08)	852 (100)
Female	196 (33.56)	187 (32.02)	276 (47.26)	287 (49.14)	112 (19.18)	110 (18.84)	584 (100)
Total	490 (34.12)	496 (34.54)	664 (46.24)	676 (47.08)	282 (19.64)	264 (18.38)	1436 (100)
Marital Status							
Single	229 (28.27)	230 (28.40)	377 (46.54)	410 (50.62)	204 (25.19)	170 (20.99)	810 (100)
Married	261 (41.70)	266 (42.50)	286 (45.69)	266 (42.50)	78 (12.46)	94 (15.02)	626 (100)
Total	490 (34.12)	496 (34.54)	664 (46.24)	676 (47.08)	282 (19.64)	264 (18.38)	1436 (100)
Education							
No formal Education	57 (59.38)	59 (61.46)	32 (33.33)	27 (28.13)	7 (7.30)	10 (10.42)	96 (100)
Upto High School	114 (45.78)	112 (44.98)	89 (35.74)	96 (38.55)	46 (18.47)	41 (16.47)	249 (100)

Bachelor degree	209 (37.86)	212 (38.48)	247 (44.75)	252 (45.65)	96 (17.39)	88 (15.94)	552 (100)
Master/Professional Degree	85 (18.48)	91 (19.78)	258 (56.09)	255 (55.43)	117 (25.43)	114 (24.78)	460 (100)
Others	25 (31.65)	22 (27.85)	38 (48.10)	46 (58.23)	16 (20.25)	11 (13.92)	79 (100)
Total	490 (34.12)	496 (34.54)	664 (46.24)	676 (47.08)	282 (19.64)	264 (18.38)	1436 (100)
Occupation							
Student	159 (24.96)	164 (25.67)	313 (49.14)	338 (53.06)	165 (25.90)	135 (21.19)	637 (100)
Home maker	98 (54.14)	88 (48.62)	68 (37.57)	72 (39.78)	15 (8.29)	21 (11.60)	181 (100)
Agri. And Allied Activities	60 (49.59)	61 (50.41)	48 (39.67)	48 (39.67)	13 (10.74)	12 (9.92)	121 (100)
Salaried entrepreneurs/ Self Business	146 (36.68)	156 (39.20)	188 (47.24)	170 (42.71)	64 (16.08)	72 (18.09)	398 (100)
Others	27 (27.27)	27 (27.27)	47 (47.47)	48 (48.48)	25 (25.77)	24 (24.24)	99 (100)
Total	490 (34.12)	496 (34.54)	664 (46.24)	676 (47.08)	282 (19.64)	264 (18.38)	1436 (100)
Annual Family Income							
Less than Rs. 1,00,000	233 (35.20)	240 (36.25)	290 (43.81)	303 (45.77)	139 (21.00)	119 (17.98)	662 (100)
Rs. 1,00,000 to less than Rs. 4,00,000	193 (40.55)	192 (40.34)	214 (44.96)	224 (47.06)	69 (14.50)	60 (12.61)	476 (100)
Rs. 4,00,000 to less than Rs. 7,00,000	43 (24.16)	48 (26.97)	90 (50.56)	85 (47.75)	45 (25.28)	45 (25.28)	178 (100)
Rs. 7,00,000 to less than Rs. 10,00,000	10 (17.24)	9 (15.52)	34 (58.62)	27 (46.55)	14 (24.14)	22 (37.93)	58 (100)
Above Rs. 10,00,000	11 (17.74)	7 (11.29)	36 (58.06)	37 (59.68)	15 (24.19)	18 (29.03)	62 (100)
Total	490 (34.12)	496 (34.54)	664 (46.24)	676 (47.08)	282 (19.64)	264 (18.38)	1436 (100)
Residence Location							
Rural	189 (30.93)	196 (32.08)	274 (44.84)	289 (47.29)	148 (24.22)	126 (20.62)	611 (100)
Urban	301 (36.48)	300 (36.36)	390 (47.27)	387 (46.91)	134 (16.24)	138 (16.72)	825 (100)
Total	490 (34.12)	496 (34.54)	664 (46.24)	676 (47.08)	282 (19.64)	264 (18.38)	1436 (100)

Division							
Bengaluru	199 (34.85)	188 (32.92)	265 (46.41)	279 (48.86)	107 (18.74)	104 (18.21)	571 (100)
Mysuru	35 (12.41)	36 (12.77)	150 (53.20)	150 (53.20)	97 (34.64)	96 (34.04)	282 (100)
Belagavi	137 (50.55)	139 (51.29)	100 (36.90)	103 (38.00)	34 (12.55)	29 (10.70)	271 (100)
Gubarga	119 (38.14)	133 (42.63)	149 (47.76)	144 (46.15)	44 (14.10)	35 (11.22)	312 (100)
Total	490 (34.12)	496 (34.54)	664 (46.24)	676 (47.08)	282 (19.64)	264 (18.38)	1436 (100)

To find out whether there is any significant variation across the demography in the awareness level on the quality of tea and coffee produced in India, analysis of variance is carried out on the calculated mean.

Table 4.22: ANOVA Single factor for variation in awareness level on quality across demography

Demography	**Calculated p value**		**Result (significant if $p < 0.05$ at 95%)**
	Tea	**Coffee**	
Age group	0.000546	0.004022	Significant for both tea and coffee
Gender	0.000171	0.002412	
Marital Status	0.079915	0.060093	Insignificant for both tea and coffee
Education	0.006941	0.006703	Significant for both tea and coffee
Occupation	0.020244	0.000495	
Annual family income	0.000129	0.002771	
Place of residence	0.029204	0.002902	
Division	0.03315	0.027736	

The awareness level on the quality of coffee and tea is very poor among the respondents in the study area and there is significant variation as per the age group, youngsters are better aware of the quality that elderly respondents.. The awareness level does not vary much as per the gender. A maximum of 21% of the 'single' respondents have complete awareness, whereas only 15% of the 'married' respondents are having complete awareness. Only 10.42% of respondents having no formal education have a complete awareness on the quality of coffee (tea: 7.30%) produced in India. The awareness level increases with the educational qualification with a high percentage of about 25% having complete awareness. There is also a strong association between the occupation and the awareness level on the quality. Student's category, with maximum of 21.19% for coffee and 25.90% for tea are completely aware of the quality. Agri & allied activities and home makers relatively have poor awareness level.

4.3. Evaluation Criteria and Importance to 'Country of Origin'

Before consumers decide whether or not they prefer a product, they decide what constitutes a good or a bad product. Another words, they should create criteria against which to measure products' ability to meet their requirements. The attitude formed by the consumers and the intention to purchase is largely influenced by the consumers' evaluation process and the criteria considered for evaluation. Therefore, effort has been made to understand the consumers' behavior in evaluating the brand and the relative importance given to the 'country of origin'.

The different criteria usually consumers consider for brand evaluation of coffee and tea were listed based on the literature and brainstorming with stakeholders. The importance of origin information to consumers when they buy coffee/tea and the extent of 'country of origin' effect compared to other aspects upon product evaluation and decision making is measured by asking the respondents to rate the various criteria such as A1: Aroma, A2: Availability, A3: Brand, A4: Country of origin,A5: Cuppage (make more cup of coffee/tea for a given quantity), A6: Ease of use (instant or ready to drink coffee, tea), A7: Flavour, A8: Information label, A9: Package, A10: Price and A11: Taste. The respondents were asked to rate these criteria in a five point scale from low important to extremely important when they buy coffee/tea.

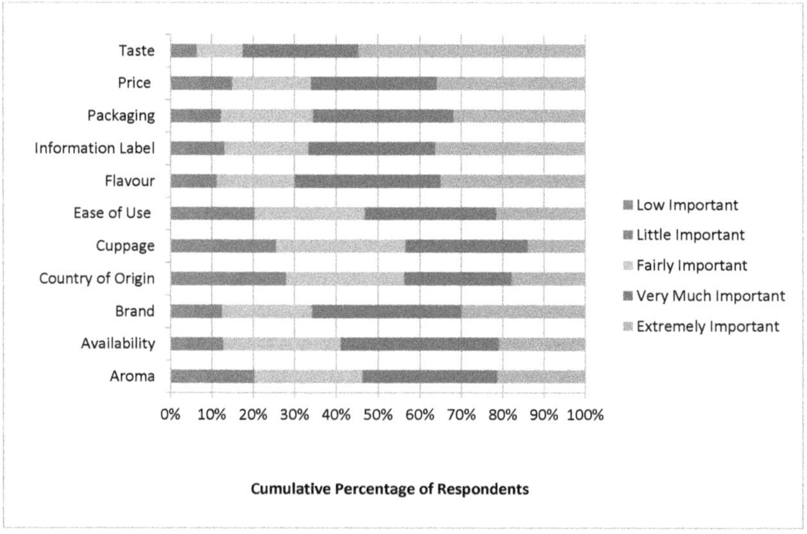

Fig 4.13: Evaluation Criteria and Relative Importance

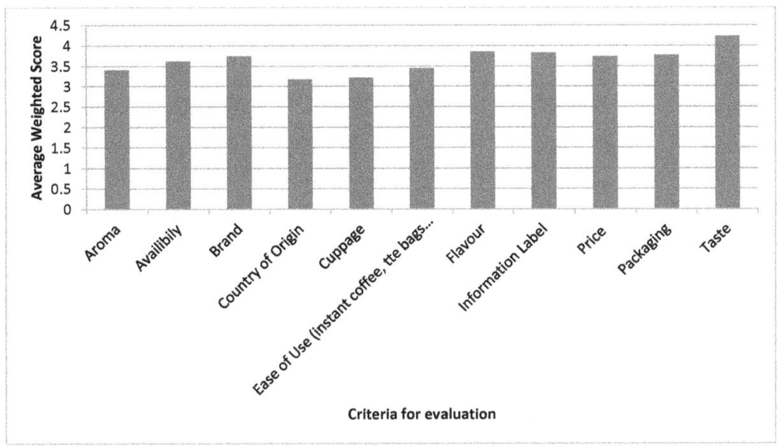

Fig. 4.14: Average Weighted Score for various Criteria

The following figure shows the percentage of respondents and their relative importance to various criteria in evaluating coffee/tea brands.

The figure shows the respondents of the study area declare that they consider not just one criterion for brand evaluation. All the listed criteria were scored in the range of 3 – 4, that is fairly important to very much important except the criterion 'Taste', which scored 4.23 (extremely important). It can be inferred that Indian consumers give utmost importance to taste. The criterion 'country of origin' has scored least among all the listed criteria. This may be because, consumers are neither aware of the places where these products are originated nor aware of the taste and quality of coffee/tea associated with the place where coffee/tea grown.

Interestingly, the attributes 'brand' and 'information label' are given the second priority by the respondents. The importance to these attribute is very much essential for promoting the brand on COOL.

To get more clarity on this, data are presented as per the demography and analyzed.

Table 4. 23: Division-wise Average Weighted Score for Evaluation Criteria

Division	Average Weighted Score for Attributes										
	A1	A2	A3	A4	A5	A6	A7	A8	A9	A10	A11
Bengaluru	3.66	3.66	3.87	3.22	3.19	3.47	4.02	3.88	4.02	3.91	4.43
Mysuru	3.52	3.60	3.79	2.85	3.17	3.47	3.93	3.79	3.55	3.71	4.19
Belagavi	3.16	3.44	3.38	3.15	3.07	3.08	3.49	3.49	3.24	3.34	3.96
Gulberga	3.29	3.76	3.93	3.49	3.44	3.76	3.97	4.17	4.15	4.12	4.33
Overall	3.41	3.62	3.74	3.18	3.22	3.44	3.85	3.83	3.74	3.77	4.23

ANOVA						
Source of Variation	SS	df	MS	F	P-value	F crit
Between Groups	3.799518	10	0.379952	5.432292	9.91E-05	2.132504
Within Groups	2.308125	33	0.069943			
Total	6.107643	43				

The above table and the ANOVA results show that there is not much significant difference in the evaluation criteria among the respondents of different divisions. Respondents in all the four divisions have rated 'taste' as the most important attribute they consider in the brand evaluation.

The average weighted score for the various attributes as per the age group of the respondents is shown in table below.

Table 4.24: Age Group -wise Average Weighted Score for Evaluation Criteria

Age group	Average Weighted Score for Attributes										
	A1	**A2**	**A3**	**A4**	**A5**	**A6**	**A7**	**A8**	**A9**	**A10**	**A11**
18-25	3.53	3.64	3.99	3.12	3.26	3.58	4.14	3.97	3.99	3.88	4.36
26-35	3.43	3.56	3.67	3.00	3.17	3.41	3.89	3.71	3.83	4.00	4.30
36-45	3.23	3.52	3.56	3.19	3.10	3.28	3.65	3.46	3.55	3.66	4.02
46-55	3.24	3.51	3.63	3.01	3.05	3.39	3.71	3.74	3.63	3.81	3.94
56-65	3.38	3.56	3.78	3.11	3.10	3.31	3.69	3.75	3.62	3.77	4.24
64 and above	3.68	3.95	3.78	3.67	3.66	3.71	4.00	4.36	3.97	3.29	4.53
Total	3.41	3.62	3.74	3.18	3.22	3.45	3.85	3.83	3.77	3.74	4.23

ANOVA						
Source of Variation	SS	df	MS	F	P-value	F crit
Between Groups	5.601761	10	0.560176	12.37121	6.94E-11	2.007792
Within Groups	2.490433	55	0.045281			
Total	8.092194	65				

Though there is not much of significant difference in the brand evaluation patter as per the age group, the elderly respondents give slightly higher importance to 'country of origin' compared to 'packaging'. Overall, the ANOVA reveals that the age group does not seem to affect the evaluation patter.

Table below shows the brand evaluation criteria and the relative importance as per the gender.

Table 4.25: Gender-wise Average Weighted Score for Evaluation Criteria

Gender Wise	Average Weighted Score for Attributes										
	A1	A2	A3	A4	A5	A6	A7	A8	A9	A10	A11
Male	3.36	3.64	3.73	3.27	3.23	3.48	3.79	3.84	3.80	3.78	4.22
Female	3.45	3.60	3.75	3.08	3.21	3.43	3.91	3.82	3.74	3.70	4.23
Over all	3.41	3.62	3.74	3.18	3.22	3.45	3.85	3.83	3.77	3.74	4.23

ANOVA

Source of Variation	SS	df	MS	F	P-value	F crit
Rows	0.0022	1	0.0022	0.632184	0.445017	4.964603
Columns	1.883582	10	0.188358	54.12591	2.33E-07	2.978237
Error	0.0348	10	0.00348			
Total	1.920582	21				

Overall, there is no significant variation; however, there is a gender related pattern for the attribute 'country of origin'. The 'male' respondents offer higher importance than the 'female' respondents.

To find out whether the relative importance and evaluation pattern varies as per marital status, the weighted scores are presented accordingly and analyzed.

Table 4.26: Marital Status-wise Average Weighted Score for Evaluation Criteria

Marital Status	Average Weighted Score for Attributes										
	A1	A2	A3	A4	A5	A6	A7	A8	A9	A10	A11
Married	3.50	3.67	3.86	3.24	3.28	3.55	4.00	3.97	3.90	3.78	4.32
Un Married	3.31	3.59	3.62	3.14	3.15	3.32	3.70	3.70	3.63	3.73	4.12
Over all	3.41	3.62	3.74	3.18	3.22	3.45	3.85	3.83	3.77	3.74	4.23

ANOVA

Source of Variation	SS	df	MS	F	P-value	F crit
Between Groups	1.883582	10	0.188358	55.99838	5.66E-08	2.853625
Within Groups	0.037	11	0.003364			
Total	1.920582	21				

There is no association found as per the marital status in the brand evaluation patter. Both, married and unmarried offer highest and almost equal weightage to 'taste' and lowest to 'country of origin'. Unmarried respondents relatively rate less for 'cuppage' compared to their married counterparts.

The following table shows the relative weighted score for the attributes as per the educational qualifications of the respondents.

Table 4.27: Education Qualification-wise Average Weighted Score for Evaluation Criteria

Education	Average Weighted Score for Attributes										
	A1	A2	A3	A4	A5	A6	A7	A8	A9	A10	A11
Up to high school	3.29	3.41	3.56	2.90	3.27	3.27	3.22	3.62	3.41	3.34	3.92
Up to high school	3.38	3.59	3.70	2.99	3.28	3.40	3.81	3.77	3.80	3.84	4.32
Bachelors degree	3.48	3.66	3.83	3.31	3.25	3.47	3.92	3.92	3.85	3.83	4.30
Masters/ professional degree	3.31	3.61	3.59	3.09	2.99	3.35	3.76	3.73	3.64	3.54	4.07
others	3.02	3.54	3.61	3.25	3.69	3.49	3.73	3.62	3.75	3.51	3.99
Overall	3.41	3.62	3.74	3.18	3.22	3.45	3.85	3.83	3.77	3.74	4.23

ANOVA

Source of Variation	SS	df	MS	F	P-value	F crit
Between Groups	3.796858	10	0.379686	11.90034	1.39E-09	2.053901
Within Groups	1.40384	44	0.031905			
Total	5.200698	54				

Overall, there is no significant difference in the evaluation pattern as per the educational qualification. However, considering the attributes 'country of origin', 'information label', educated respondents have given higher importance than respondents with no formal education.

Table 4.28: Occupation-wise Average Weighted Score for Evaluation Criteria

Occupation	Average Weighted Score for Attributes										
	A1	A2	A3	A4	A5	A6	A7	A8	A9	A10	A11
Student	3.57	3.70	3.96	3.28	3.34	3.56	4.05	4.01	3.92	3.81	4.35
Home maker	3.21	3.50	3.56	3.10	3.18	3.37	3.74	3.74	3.68	3.85	4.10
Agri and Allied Activities	3.18	3.49	3.38	3.14	2.96	3.15	3.32	3.27	3.50	3.54	4.07
Salaried entrepreneurs/ Self Business	3.30	3.65	3.63	3.26	3.21	3.39	3.80	3.76	3.66	3.73	4.12
Others	3.53	3.81	3.86	2.88	3.13	3.59	3.97	3.99	3.95	3.68	4.26
Overall	3.41	3.62	3.74	3.18	3.22	3.45	3.85	3.83	3.77	3.74	4.23

ANOVA						
Source of Variation	SS	df	MS	F	P-value	F crit
Between Groups	4.679531	10	0.467953	12.36158	7.83E-10	2.053901
Within Groups	1.66564	44	0.037855			
Total	6.345171	54				

The average weighted score for various attributes in evaluating the brand as per annual family income is presented in the table below and analyzed.

Table 4.29: Annual Family Income-wise Average Weighted Score for Evaluation Criteria

Annual family Income	Average Weighted Score for Attributes										
	A1	A2	A3	A4	A5	A6	A7	A8	A9	A10	A11
Less than Rs. 1,00,000	3.53	3.70	3.94	3.30	3.36	3.55	4.04	4.01	3.91	3.83	4.35
Rs. 1,00,000 to less than Rs. 4,00,000	3.23	3.53	3.59	3.07	3.15	3.34	3.70	3.69	3.68	3.82	4.07
Rs. 4,00,000 to less than Rs. 7,00,000	3.26	3.61	3.54	3.31	3.12	3.31	3.48	3.43	3.54	3.57	4.08
Rs. 7,00,000 to less than Rs. 10,00,000	3.31	3.61	3.59	3.14	3.05	3.33	3.80	3.74	3.65	3.68	4.16
Above Rs. 10,00,000	3.56	3.77	3.86	2.92	3.11	3.62	3.99	3.93	3.92	3.54	4.22
Overall	3.41	3.62	3.74	3.18	3.22	3.45	3.85	3.83	3.77	3.74	4.23

ANOVA						
Source of Variation	SS	df	MS	F	P-value	F crit
Between Groups	4.580284	10	0.458028	17.34717	3.42E-12	2.053901
Within Groups	1.16176	44	0.026404			
Total	5.742044	54				

Even the annual family income do not have an overall association with the brand evaluation. Attribute 'taste' remains to be the highest important criteria among all income class respondents.

Finally, the weighted scores for various criteria are presented as per the place of residence of respondents and analyzed.

Table 4.30: Place of Residence-wise Average Weighted Score for Evaluation Criteria

Residence Location	Average Weighted Score										
	A1	A2	A3	A4	A5	A6	A7	A8	A9	A10	A11
Rural	3.33	3.59	3.77	3.20	3.23	3.45	3.83	3.87	3.79	3.76	4.19
Urban	3.46	3.62	3.71	3.16	3.21	3.45	3.86	3.82	3.75	3.73	4.26
Over all	3.41	3.62	3.74	3.18	3.22	3.45	3.85	3.83	3.77	3.74	4.23

ANOVA

Source of Variation	SS	df	MS	F	P-value	F crit
Between Groups	1.899027	10	0.189903	122.1596	8.51E-10	2.853625
Within Groups	0.0171	11	0.001555			
Total	1.916127	21				

The respondents belonging to the urban and rural area, both seem to have overall similar evaluation pattern. Individually also, the score calculated for each attributes does not significantly vary as per their place of residence.

4.4. Consumer Attitude towards foreign brands of Coffee/Tea

To get the clarity on the possibility of promoting plantation commodities on 'country of origin labelling', it is also required to assess the Indian consumers' attitude towards foreign brands of these commodities. The unfavourable attitude towards foreign brands will result into intention to purchase domestic brand and favour the promotion of domestic brands on 'country of origin labelling'.

The Belief Importance model (B-I model) has been used to measure the attitude of consumers towards foreign brands. The B-I model allows the comparison of affective responses toward competing brands and the formula adopted is given below:

$$A_o = \sum_{i=1}^{m} B_{io} I_i$$

where,

A_o = Attitude toward foreign brand

B_{io} = Belief that foreign brand does well or poorly when its attribute (i) is compared with those of domestic brands

I_i = Importance of attribute (i) in selecting the brand

i = attribute 1, 2, … m

Five attributes were identified and two sets of statements, one for belief and another for importance were developed and the respondents were asked to choose from strongly disagree to strongly agree in a five point scale.

The attributes such as quality, price, available variety, social status and promotional schemes were identified and respondents were asked to rate how far they believe that these attributes are present in the foreign brands and then rate how much importance they give to these attributes for brand choice. Ideally, the score will vary from -20 to +20. Positive score reflects the consumers' favourable attitude, the negative score reflects unfavourable attitude towards foreign brands. In a similar manner, the intensity of favourableness or unfavourableness will be decided by the score itself. Higher the score (leaving the sign), higher will be the intensity, vice versa.

It is derived from the study that the average score calculated as + 2.73, which means, though the score is very negligible the consumers have a favourable attitude towards foreign brands of coffee/tea. Most of the respondents had a belief that the use of foreign brands elevates their social status.

To find out whether the attitude varies as per the demography, the data are presented accordingly and analyzed. The following table shows the detail.

Table 4.31: Attitude Score for the Foreign Brands

Age group-wise	$\Sigma Bi \times Ii$	**Gender-wise**	$\Sigma Bi \times Ii$	**Education qualifications-wise**	$\Sigma Bi \times Ii$
18-25	2.74	Male	2.56	no formal education	3.64
26-35	2.84	Female	2.89	Upto high school	2.83
36-45	3.61	Total	2.73	Bachelor's degree	2.10
46-55	3.41	Marital Status		Master's/ professional degree	2.54
56-65	2.69	Married	2.46	others	2.56
66 and above	1.10	Un- Married	3.00	Total	2.73
Total	2.73	Total	2.73		
Occupation-wise		**Annual family Income-wise**	2.48	Place of Residence-wise	2.88
Student	2.60	Less than Rs. 1,00,000	3.23	Rural	2.58
Home Maker	2.61	Rs. 1,00,000 to less than Rs. 4,00,000	2.51	Urban	2.73
Agri. and Allied Activities	3.34	Rs. 4,00,000 to less than Rs. 7,00,000	2.69	**Total**	

Salaried entrepreneurs/ Self Business	2.66	Rs. 7,00,000 to less than Rs. 10,00,000	2.76	**Division-wise**	
Others	2.46	Above Rs. 10,00,000	2.73	Bengaluru	2.64
Total	2.73	**Total**		Mysuru	2.12
Place of Residence-wise				Belagavi	3.81
Rural	2.88			Gulberga	2.33
Urban	2.58			**Total**	2.73
Total	2.73				

There is a strong gender related attitude towards foreign brands. The relative attitude of respondents in the middle age group, ie. 36 to 55, the degree of favourableness is higher and the degree is lowest among above 56 age group. There is also a gender related pattern, female respondents found to have higher degree of favourableness than males. Similar is the case with unmarried respondents. It can also be inferred from the table that respondents with no formal education relatively favour more to foreign brands and the degree of favourableness reduces with advancement in education. As far the division-wise score is concerned, the respondents of Belagavi region has the highest degree of favourabless of +3.81 and Mysuru with lowest of +2.12.

4.5. The Consumer Ethnocentrism

Several studies have been conducted across the globe and their research findings revealed that the consumer ethnocentric tendencies will have an influence on attitude formation on domestic brand and the purchase intention. The studies also revealed that consumers with high ethnocentrism will prefer domestic brand over foreign. Therefore, Consumer ethnocentric tendency scale (CETSCALE) developed by Shimp & Sharma (1987), was used to measure the respondents' ethnocentric tendency when it comes to the purchase of coffee/tea. The following table shows the detail.

A total of 17 statements with the highest rating of each statement as 5, the maximum weighted score for the Consumer Ethnocentrism may be 85. It is revealed from the mean CET score for the consumers in the study area is 62.73, which is at a moderate level.

Further, Factor Analysis is done using principal component analysis (PCA) and a varimax rotation method using MINITAB to identify the main aspects of consumer ethnocentric tendencies in consumers. The individual items of CETSCALE were grouped into four factors. According to the PCA analysis, each item of CETSCALE was loaded to four different factors and the total variance of the four factor solutions for CE was 10.7285 and the percentage of variance was 63.10%. The table below shows the extracted factor loading.

Table 4.32: Rotated Factor Loadings and Communalities for Consumer Ethnocentrism

Variable	Factor1	Factor2	Factor3	Factor4
1. Indian People should always	0.686	-0.131	0.272	-0.197
2. Only those products that are	0.658	-0.345	-0.034	-0.152
3. Buy Indian - Made Products,	0.793	-0.168	0.209	-0.090
4. Indian Products, first, last	0.726	-0.144	0.190	-0.268
5. Purchasing foreign made coff	0.163	-0.263	0.111	-0.792
6. It is not right to purchase	0.240	-0.180	0.229	-0.680
7. A real Indian should always	0.327	-0.206	0.519	-0.512
8. We should purchase tea/ coff	0.451	-0.359	0.440	-0.250
9. It is always best to purchas	0.481	-0.226	0.600	-0.129
10. There should be very little	0.443	-0.590	0.096	-0.194
11. Indian should not buy forei	0.087	-0.169	0.694	-0.191
12. Curbs should be put on all	0.230	-0.681	0.122	-0.322
13. It may cost in the long run	0.456	-0.406	0.492	-0.037
14. Foreigners should not be al	0.079	-0.632	0.247	-0.422
15. Foreigner tea/ coffee shoul	0.118	-0.619	0.421	-0.186
16. We should obtain from forei	0.417	-0.665	0.179	0.011
17. Indian consumers who purcha	0.153	-0.556	0.444	-0.294

Source: Based on survey data, extracted using PCA Varimax Rotation, MINITAB

The seventeen variables are grouped as per their coefficients and named. The following table shows the detail.

Table 4.33: Factors and Rotated Component Matrix for Consumer Ethnocentrism

Factor No.	Factor Name	Statement	Component			
			1	2	3	4
F1	Passion for Indian Products	CE5: Purchasing foreign made tea/ coffee is un-Indian	0.736			
		CE3: Buy Indian- made products, keep India working.	0.709			
		CE7: A real Indian should always buy India-made tea/ coffee	0.680			
		CE9: It is always best to purchase Indian tea/ coffee	0.660			

		CE4: Indian products, first, last, and foremost	0.656			
F2	Entry Restriction	CE16: We should obtain from foreign countries only those products that we cannot obtain in our country		0.649		
		CE14: Foreigners should not be allowed to put their tea/ coffee on your market		0.645		
		CE12: Curbs should be put on all imports		0.635		
F3	Economic Instability	CE17: Indian consumers who purchase products made in other countries are responsible for putting their fellow Indians out of work.			0.617	
		CE13: It may cost me in the long run but I prefer to support Indian tea/ coffee			0.617	
		CE15: Foreign tea/ coffee should be taxed heavily to reduce their entry into India			0.609	
		CE6: It is not right to purchase foreign tea/ coffee because it puts Indian out of job			0.604	
F4	Consumer Animosity	CE1: Indian people should always buy Indian tea/ coffee instead of imported tea/ coffee.				0.601
		CE10: There should be very little trading or purchasing of tea/ coffee from other countries unless out of necessity				0.590
		CE8: We should purchase tea/ coffee manufactured in India instead of letting other countries get rich off us				0.589
		CE2: Only those products that are unavailable in India should be imported.				0.576

	CE11: Indian should not buy foreign tea/ coffee because it hurts Indian business and causes unemployment	0.554

Source: Based on survey data, extracted using PCA Varimax Rotation, MINITAB

Factor 1: The 5 variables in the CETSCALE viz. CE5, CE3, CE7, CE9 and CE4 having high coefficient were grouped under Factor 1. Each variable reflects the consumers' love for his country and passion for their country's products. This group of variables is labeled as 'Passion for Indian Products'. This behavior exhibited by the respondents in the study area is a positive one for the marketers and policy makers to promote coffee/tea on COOL.

Factor 2: The three variables CE16, CE14 and CE12 are the next set of variables having high coefficient and they are grouped under Factor 2. These variables expresses the respondents concern on the import of foreign brands in the Indian market. The respondents in the study area feel that such brands should be made available in Indian market, only when we cannot obtain these products produced in our country. Thus the factor is named as 'Entry Restriction'.

Factor 3: Another set of four variables CE17, CE13, CE15 and CE6 share some commonalities and they are clubbed together and labeled as 'Economic Instability'. These set of variables reflects the consumers' apprehension about the entry of foreign brands on Indian economy. Respondents were of the opinion that the entry of foreign brands will result into negative impact on employment and economy. However, it is ranked as only as a third factor of their concern.

Factor 4: The reaming 5 set of variables CE1, CE10, CE8, CE2, and CE11 were with a least coefficient and grouped under Factor 4. This set of variables mostly reflect the consumers' emotions like fear, hate towards foreign products. Thus, the Factor4 is named as 'Consumer Animosity'.

Accordingly, the four factors as per the order of high priority to low priority viz. Passion for Indian Products, Entry Restriction, Economic Instability and Consumer Animosity determines the respondents CET.

To find out, whether the consumer ethnocentric tendency (CET) varies as per the demography, the weighted scores are presented as per the demography and analysed. The following table shows the average weighted score for each of the statements as per the demography.

Division-wise	Average Weighted Score for Statements																	Total
	1	2	3	4	5	6	7	8	9	10	11	12	13	14	15	16	17	
Division																		
Bengaluru	3.92	3.90	4.25	4.06	3.65	3.77	3.93	4.02	3.31	3.81	3.62	3.92	3.78	3.74	3.89	3.52	3.69	64.78
Mysuru	3.54	3.73	3.99	3.75	3.15	3.32	3.43	3.68	2.93	3.39	3.23	3.65	3.25	3.39	3.57	3.10	3.35	58.43
Belagavi	3.39	3.50	3.75	3.52	3.20	3.35	3.52	3.55	3.07	3.34	3.20	3.57	3.34	3.45	3.54	3.24	3.41	57.93
Gulberga	3.88	4.05	4.25	4.29	4.03	4.21	4.17	4.23	3.99	4.10	4.19	4.13	4.02	4.04	4.12	3.94	4.10	69.73
Overall	3.68	3.79	4.06	3.91	3.50	3.66	3.76	3.87	3.33	3.66	3.56	3.82	3.60	3.66	3.78	3.45	3.64	62.72
Age Group																		
18-25	3.22	3.50	3.97	4.00	3.58	4.04	4.36	4.71	4.98	4.97	5.10	5.28	5.73	5.59	6.01	6.30	6.31	81.64
26-35	3.73	3.88	4.02	3.93	3.47	3.49	3.79	3.74	3.84	3.70	3.72	3.67	3.79	3.48	3.53	3.82	3.78	63.35
36-45	3.52	3.59	3.81	3.70	3.36	3.38	3.59	3.54	3.62	3.58	3.41	3.35	3.64	3.31	3.54	3.55	3.54	60.00
46-55	3.52	3.67	3.99	3.89	3.36	3.41	3.55	3.69	3.89	3.65	3.49	3.51	3.92	3.38	3.60	3.74	3.50	61.73
56-65	3.91	3.81	4.06	4.03	3.55	3.87	3.80	3.85	4.07	3.61	3.61	3.68	3.86	3.70	3.86	3.80	3.94	64.97
66 and above	3.15	2.25	3.06	2.75	2.61	2.54	2.88	2.92	3.06	2.83	2.69	2.46	2.77	2.85	2.44	2.90	2.75	46.89
Total	3.68	3.80	4.06	3.91	3.33	3.51	3.66	3.76	3.87	3.66	3.60	3.56	3.82	3.45	3.66	3.78	3.64	62.73
Gender Wise																		
Male	3.67	3.81	4.04	3.91	3.32	3.54	3.65	3.76	3.87	3.69	3.57	3.56	3.83	3.46	3.71	3.74	3.61	62.72
Female	3.71	3.78	4.09	3.89	3.32	3.47	3.67	3.76	3.88	3.64	3.63	3.55	3.80	3.40	3.57	3.81	3.66	62.60
Over all	3.68	3.80	4.06	3.91	3.33	3.51	3.66	3.76	3.87	3.66	3.60	3.56	3.82	3.45	3.66	3.78	3.64	62.73
Marital Status																		
Married	3.76	3.88	4.21	3.99	3.27	3.56	3.69	3.85	3.97	3.72	3.63	3.61	3.91	3.49	3.75	3.87	3.64	63.77

Un- Married	3.63	3.72	3.90	3.85	3.41	3.47	3.68	3.69	3.79	3.62	3.56	3.52	3.74	3.42	3.56	3.69	3.65	61.88
0ver all	3.68	3.80	4.06	3.91	3.33	3.51	3.66	3.76	3.87	3.66	3.60	3.56	3.82	3.45	3.66	3.78	3.64	62.73
Educational Qualifications																		
no formal education	3.6425	3.7125	3.63	3.585	3.5975	3.6325	3.53	3.605	3.635	3.555	3.6725	3.4725	3.59	3.035	3.5525	3.7275	3.64	60.82
Upto high school	3.73	3.8225	3.9975	3.9525	3.6325	3.6725	3.8675	3.79	3.9	3.7425	3.695	3.765	3.7875	3.7625	3.775	3.8325	3.875	64.60
Bachelor's degree	3.8425	3.9375	4.155	4.0525	3.385	3.6225	3.805	3.9025	4.0175	3.7675	3.675	3.6875	3.99	3.5225	3.785	3.945	3.7375	64.83
Master's/ profession-al degree	3.505	3.6	3.9575	3.7175	3.0675	3.23	3.42	3.55	3.7525	3.4725	3.4725	3.2475	3.725	3.165	3.45	3.5	3.385	59.22
others	3.5775	3.71	4.0175	3.7175	3.2575	3.23	3.485	3.085	4.135	3.3625	3.9325	3.33	4.04	3.65	3.96	3.55	3.5575	61.60
Over all	3.68	3.80	4.06	3.91	3.33	3.51	3.66	3.76	3.87	3.66	3.60	3.56	3.82	3.45	3.66	3.78	3.64	62.73
Occupation																		
Student	3.7325	3.8475	4.23	4	3.2275	3.54	3.6725	3.875	3.9775	3.71	3.61	3.6025	3.9075	3.475	3.755	3.8675	3.6	63.63
Home Maker	3.69	3.8475	3.9875	3.905	3.5775	3.5825	3.84	3.7075	3.805	3.85	3.5825	3.77	3.7625	3.55	3.6125	3.865	3.7425	63.68
Agri. and Allied Activities	3.6675	3.5825	3.8275	3.7525	3.3725	3.4175	3.6725	3.7175	3.705	3.6925	3.5825	3.57	3.78	3.3975	3.69	3.7975	3.5725	61.80
Salaried entrepre-neurs/ Self Business	3.085	3.8175	3.9175	3.93	3.695	3.5225	3.6525	3.7675	3.845	3.7425	3.635	3.5525	3.7025	3.5925	3.5575	3.635	3.7525	62.40
Others	3.9225	4.0625	4.195	3.895	3.12	3.32	3.665	3.545	3.8625	3.6125	3.83	3.305	4.1075	3.3625	3.6525	3.99	3.645	63.09

Overall	3.68	3.80	4.06	3.91	3.33	3.51	3.66	3.76	3.87	3.66	3.60	3.56	3.82	3.45	3.66	3.78	3.64	62.73
Annual family income																		
Less than Rs. 1,00,000	3.7925	3.885	4.19	3.965	3.4025	3.61	3.76	3.895	3.9675	3.7575	3.7075	3.665	3.9425	3.5675	3.785	3.9025	3.7	64.50
Rs. 1,00,000 to less than Rs. 4,00,000	3.645	3.775	4.01	3.8975	3.35	3.5275	3.6675	3.76	3.86	3.595	3.565	3.6025	3.805	3.52	3.635	3.81	3.69	62.72
Rs. 4,00,000 to less than Rs. 7,00,000	3.715	3.825	3.975	3.945	3.155	3.385	3.52	3.62	3.85	3.6825	3.66	3.2775	3.78	3.225	3.52	3.705	3.61	61.45
Rs. 7,00,000 to less than Rs. 10,00,000	3.7425	4.135	4.365	4.225	2.7025	3.72	3.555	3.8875	4.275	3.63	2.9075	3.0575	3.755	2.665	3.0875	3.3925	2.8975	60.00
Above Rs. 10,00,000	3.325	3.1425	3.8825	3.495	3.0375	3.4225	3.55	3.395	3.665	3.1125	3.06	2.9025	3.355	2.825	3.1375	3.3375	3.4625	56.11
Overall	3.68	3.80	4.06	3.91	3.33	3.51	3.66	3.76	3.87	3.66	3.60	3.56	3.82	3.45	3.66	3.78	3.64	62.73
Residential Area																		
Rural	3.7025	3.82	4.11	3.9475	3.335	3.585	3.8	3.88	3.9125	3.7225	3.67	3.645	3.875	3.5225	3.765	3.82	3.67	63.78
Urban	3.6625	3.7675	4.0075	3.8725	3.3025	3.44	3.5375	3.675	3.83	3.61	3.5425	3.4825	3.76	3.3725	3.57	3.7475	3.6075	61.79
Over all	3.68	3.80	4.06	3.91	3.33	3.51	3.66	3.76	3.87	3.66	3.60	3.56	3.82	3.45	3.66	3.78	3.64	62.73

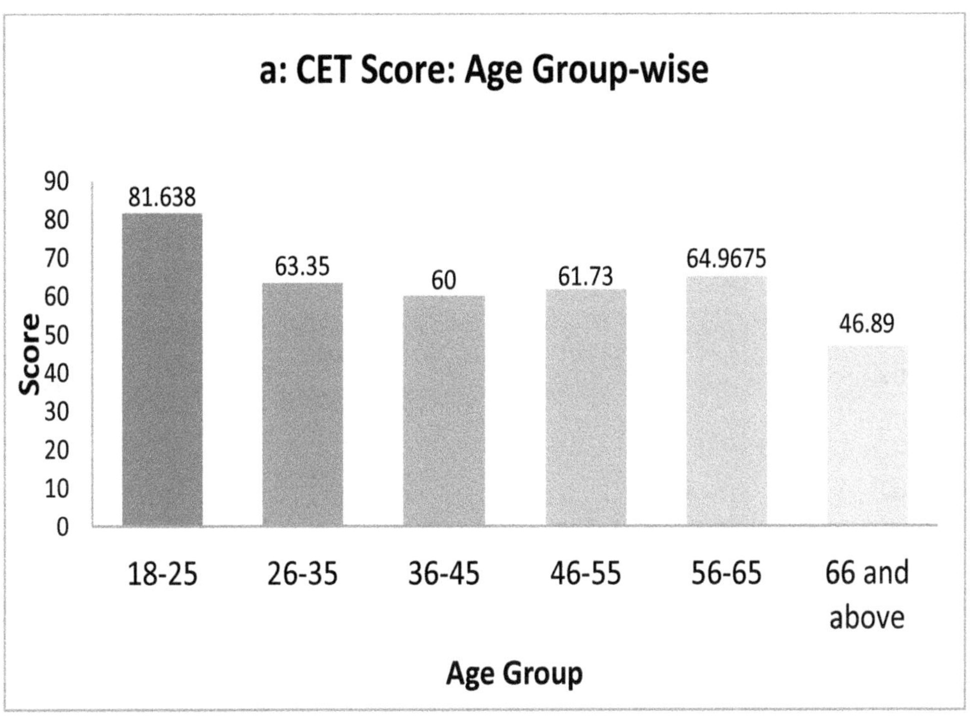
a: CET Score: Age Group-wise
Score
90
80
70
60
50
40
30
20
10
0
81.638
63.35
60
61.73
64.9675
46.89
18-25
26-35
36-45
46-55
56-65
66 and above
Age Group

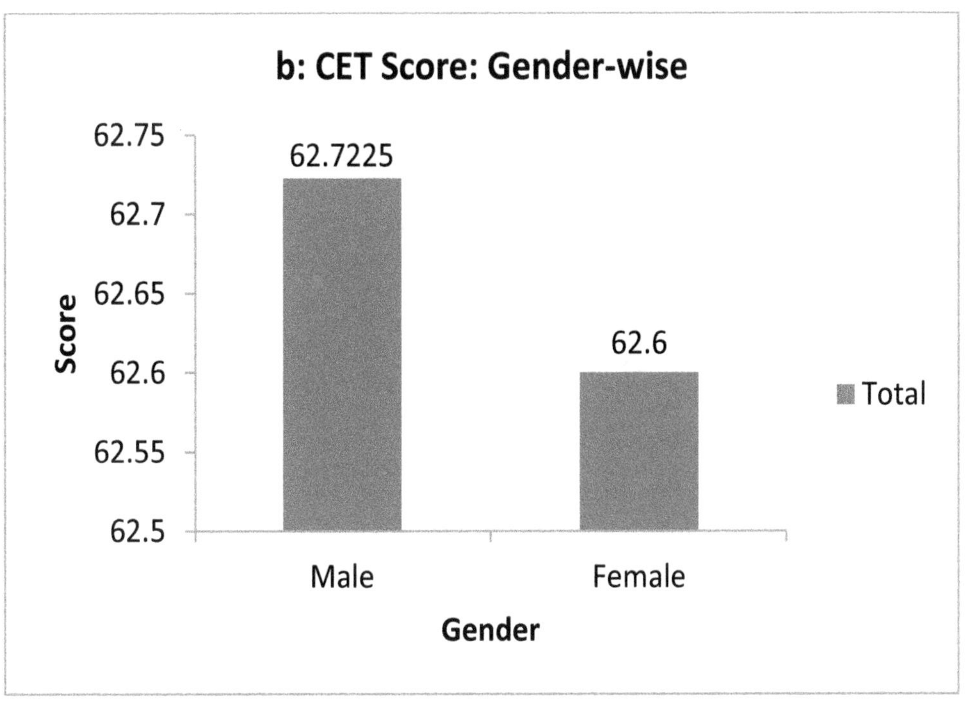
b: CET Score: Gender-wise
Score
62.75
62.7
62.65
62.6
62.55
62.5
62.7225
62.6
Total
Male
Female
Gender

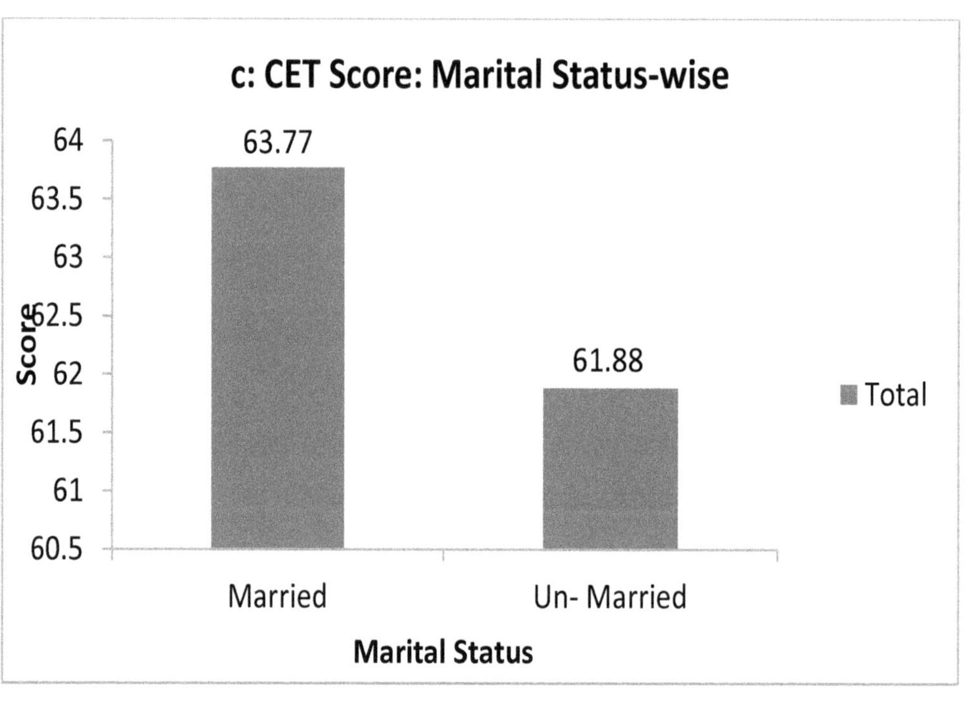
c: CET Score: Marital Status-wise
64
63.5
63
62.5
62
61.5
61
60.5
Score
63.77
61.88
Married
Un- Married
Marital Status
Total

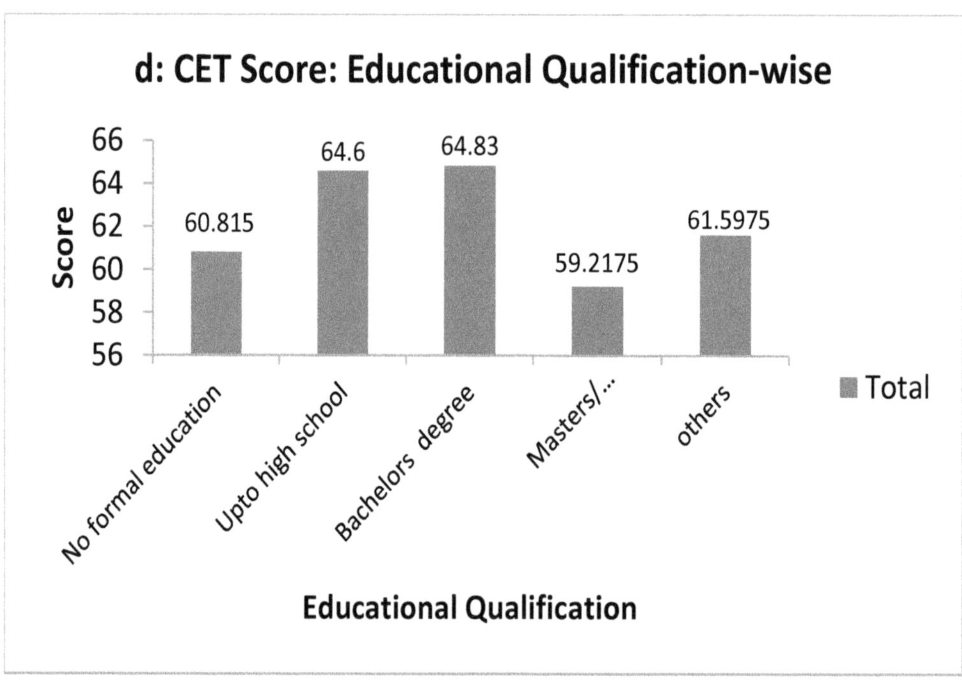
d: CET Score: Educational Qualification-wise
66
64
62
60
58
56
Score
60.815
64.6
64.83
59.2175
61.5975
No formal education
Upto high school
Bachelors degree
Masters/...
others
Educational Qualification
Total

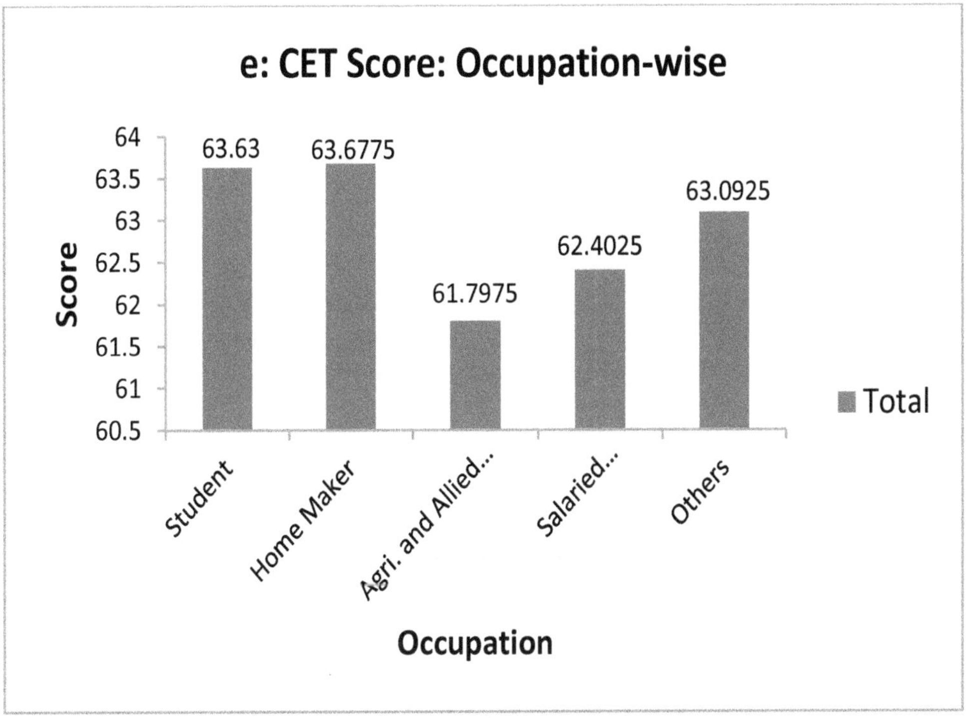

e: CET Score: Occupation-wise
Score
64
63.5
63
62.5
62
61.5
61
60.5
63.63
63.6775
61.7975
62.4025
63.0925
Student
Home Maker
Agri. and Allied...
Salaried...
Others
Occupation
Total

f: CET Score: Annual Family Income-wise
Score
66
64
62
60
58
56
54
52
50
64.495
62.715
61.45
60
56.1075
Less than Rs. 1,00,000
Rs. 1,00,000 to less than Rs. 4,00,000
Rs. 4,00,000 to less than Rs. 7,00,000
Rs. 7,00,000 to less than Rs. 10,00,000
Above Rs. 10,00,000
Annual Family Income
Total

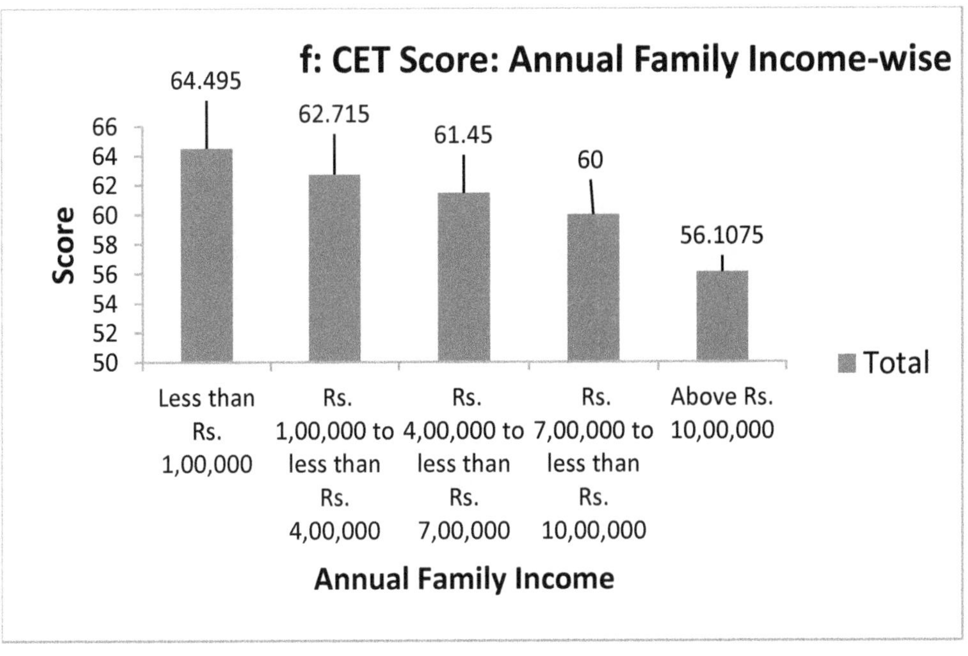

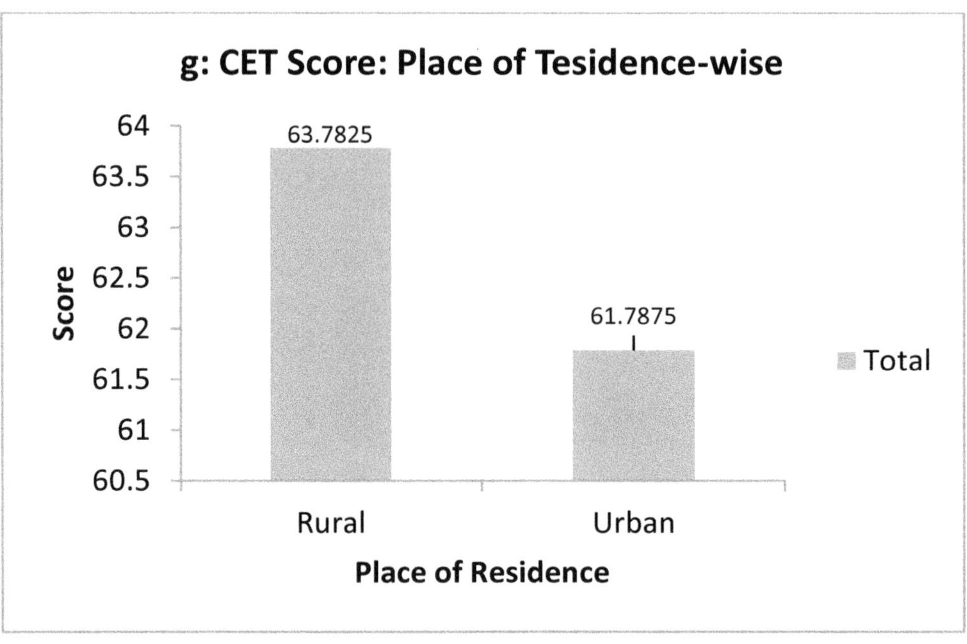

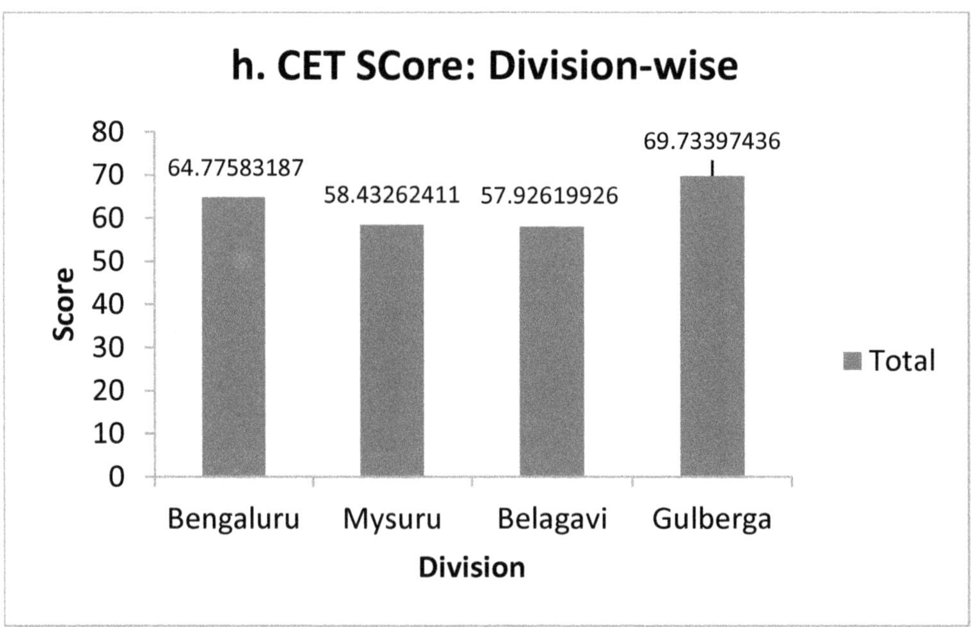

4.6. Country Image and Consumers' Perception

The understanding on the perceived image of the consumers on India as a country is of great relevance for policy makers and all the stakeholders in the coffee/tea sector before initiating the promotional activities on COOL. Nation branding plays a significant role in addition to the positive characteristics of the products. Therefore the researcher has made an attempt to assess the level of perceived country image (CI) among the respondents in the study area. A Gallup (2000) scale was adopted to measure the country image. The scale measures the evaluation of a given country (internal image) with a 24-item 4-point scale ("strongly disagree", "disagree", "agree", "strongly agree").

A total of 23 statements with the highest rating of each statement as 4, the maximum weighted score for the Country Image (CI) may be 92. It is revealed from the mean CI score for the consumers in the study area is 70.22, which is at a moderate level.

Further, Factor Analysis is done using principal component analysis (PCA) and a varimax rotation method using MINITAB to identify the main aspects of country image in consumers. The individual items of CI scale were grouped into five factors. According to the PCA analysis, each item of CI was loaded to four different factors and the total variance of the four factor solutions for CI was 13.0006and the percentage of variance was 56.50 %. The table below shows the extracted factor loading.

Table 4.35: Rotated Factor Loadings and Communalities for Country Image

	Variable	Factorl	Factor2	Factor3	Factor4	Factor5
1.	India is a successful countr	0.287	0.363	0.488	-0.238	0.009
2.	India is a country with lot	0.637	0.315	0.264	-0.022	-0.011
3.	India is a country of fair a	0.126	0.181	0.763	-0.105	0.010
4.	India is a educated and civi	0.176	0.239	0.651	-0.186	-0.110
5.	India is a country of social	0.119	0.098	0.687	-0.237	-0.320
6.	India is a country of human	0.342	0.052	0.502	-0.210	-0.313
7.	India is a country with a gl	0.680	0.150	0.201	0.019	-0.118
8.	India is a much suffered cou	0.303	0.068	0.076	-0.084	-0.702
9.	India is a country of low an	0.121	0.219	0.486	-0.104	-0.480
10.	India is a country of entre	0.206	0.450	0.133	-0.078	-0.438

11.	India is a depressed, pessi	-0.177	0.192	0.130	-0.193	-0.668
12.	India is a hardworking coun	0.164	0.629	0.265	-0.036	-0.145
13.	India is a country with sol	0.217	0.473	0.225	-0.134	-0.342
14.	India is a country with gre	0.239	0.668	0.139	-0.273	-0.111
15.	India is a country with gre	0.269	0.669	0.099	-0.236	-0.119
16.	India is a country with gre	0.756	0.283	0.128	-0.021	0.010
17.	India is a country with gre	0.108	0.564	0.168	-0.425	-0.073
18.	India is a Democratic Count	0.701	0.103	0.159	-0.206	-0.122
19.	India is a decent, clean co	-0.188	0.171	0.277	-0.677	-0.208
20.	India is a county with r	0.637	0.151	-0.000	-0.363	-0.130
21.	India is a cheerful country	0.301	0.167	0.163	-0.613	-0.116
22.	India is a development coun	0.016	0.297	0.253	-0.627	-0.161
23.	India is a fast-developing	0.327	0.106	0.080	-0.638	-0.028

Source: Based on survey data, extracted using PCA Varimax Rotation, MINITAB

The twenty three variables are grouped as per their coefficients and named. The following table shows the detail.

Table 4.36: Factors and Rotated Component Matrix for Country Image

Factor No	Factor Name	Statements	Components				
		India is a	1	2	3	4	5
F1	Cultural	CI16: Country with great culture	0.668				
		CI5: Country of social justice	0.653				
		CI19: Decent, clean country	0.644				

		CI3: Country of fair and honest people	0.641				
F2	Achievements	CI14: Country with great sports achievements		0.610			
		CI8: Much-suffered country		0.602			
		CI15: Country with great scientific achievements		0.600			
F3	People	CI18: Democratic country			0.584		
		CI20: Country rich in beautiful landscapes			0.578		
		CI2: Country with a lot of talent			0.575		
		CI22: Developed country			0.571		
		CI11: Depressed, pessimistic country			0.568		
		CI4: Educated, civilized country			0.558		
F4	Governance	CI17: Country with great economic achievements				0.545	
		CI7: Country with a glorious and rich history				0.540	
		CI9: Country of law and order				0.540	
		CI21: Cheerful country				0.534	
		CI23: Fast-developing country				0.533	
F5	Socio-economic'	CI12: Hard-working country					0.515
		CI6: Country of human freedom					0.514

	CI1: Successful country	0.509
	CI10: Country of entrepreneurs	0.461
	CI13: Country with solidarity	0.457

Source: Source: Based on survey data, extracted using PCA Varimax Rotation, MINITAB

Factor 1: The 4 variables in the CI scale viz CI 16, CI 5, CI 19 and CI 3 having high coefficient were grouped under Factor 1. Each variables reflect on the respondents endorsement of the strong traditional Indian culture. This group of variables is labeled as 'culture'.

Factor 2: The three variables CI 14, CI 8 and CI 15 are the next set of variables having high coefficient and they are grouped under Factor 2. These variables express the respondents view on the level of achievement India has made.

Factor 3: Another set of 6 variables, CI 18, CI 20, CI 2, CI 22, CI 11, and CI 4 have almost similar characteristics and they are named as 'people'. This tells about the respondents view on the people in the country, their education and talent.

Factor 4: The variables such as country's economics achievement, richness, law and order and the development activities are found to be related and grouped.The variables are CI 17, CI 7, CI 9, CI 21 and CI 23. Since these variables are very much determined by the political system, the factor 4 is labeled as 'Governance'.

Factor 5: The rest of the variables, CI 12, CI 6, CI 1, CI 10 and CI 13 are grouped as factor 5 and named as 'Socio-economic'. It outcome suggests that the factor is given lowest score. Most of the respondents feel that India lack solidarity among the citizens. They are not very confident to express that India is a successful country.

Accordingly, the five factors as per the order of high priority to low priority viz. culture, Achievements, People, Governance and Soci-economic are the major factors that determines the country image.

To find out, whether the Country Image (CI) varies as per the demography, the weighted scores are presented as per the demography and analysed. The following table shows the average weighted score for each of the statements as per the demography

Table 4. Average Weighted Score for Country Image as per Demography

	Average Weighted Score for Statements																							Total
	1	2	3	4	5	6	7	8	9	10	11	12	13	14	15	16	17	18	19	20	21	22	23	
Divisions																								
Bengaluru	3.23	3.64	3.01	3.03	3.24	3.53	3.04	3.12	3.16	2.62	3.23	2.97	3.10	3.37	3.57	3.18	3.43	3.20	3.38	3.11	2.82	2.65	3.20	72.81
Mysuru	2.91	3.30	2.86	2.82	2.94	3.30	2.91	2.88	2.72	2.39	3.03	2.79	2.78	3.06	3.34	2.88	3.17	2.95	3.13	2.83	2.48	2.37	2.90	66.75
Belagavi	2.80	3.14	2.77	2.71	2.88	3.08	2.91	2.87	2.84	2.54	2.90	2.82	2.63	2.94	2.95	2.81	2.92	3.04	2.90	2.87	2.61	2.51	2.93	65.37
Gulberga	3.27	3.56	2.99	3.21	3.36	3.43	3.37	3.16	3.33	3.03	3.37	3.23	3.22	3.46	3.49	3.29	3.41	3.42	3.41	3.26	3.26	3.03	3.36	75.91
Overall	3.05	3.41	2.91	2.94	3.11	3.33	3.06	3.01	3.01	2.65	3.13	2.95	2.93	3.21	3.34	3.04	3.23	3.15	3.20	3.02	2.79	2.64	3.10	70.22
Age Group																								
18-25	3.07	3.51	2.99	3.10	3.05	3.18	3.43	3.16	3.06	3.08	2.67	3.17	3.20	3.08	3.30	3.51	3.10	3.35	2.65	3.29	3.07	2.81	3.08	71.86
26-35	3.07	3.39	2.97	3.01	2.98	3.07	3.41	2.99	2.88	3.01	2.67	3.13	3.18	2.89	3.19	3.35	3.03	3.28	2.71	3.24	3.03	2.80	3.21	70.45
36-45	2.94	3.21	2.80	2.79	2.73	3.03	3.16	2.97	2.81	2.89	2.53	3.09	3.07	2.68	2.97	3.11	2.88	3.09	2.56	3.05	2.92	2.68	3.04	66.95
46-55	3.01	3.32	2.69	2.89	2.80	3.08	3.29	2.92	2.83	2.95	2.56	3.04	3.10	2.65	3.15	3.18	2.89	3.11	2.68	3.08	3.02	2.78	3.14	68.13
56-65	3.31	3.51	2.86	3.08	3.07	3.16	3.30	3.02	3.04	2.97	2.76	3.15	3.06	2.81	3.08	3.06	2.97	3.16	2.46	3.34	3.18	3.15	3.28	70.73
66 and above	2.46	2.67	2.23	2.46	2.17	2.44	2.56	2.46	2.42	2.40	2.27	2.46	2.69	2.48	2.54	2.27	2.31	2.31	1.90	2.38	2.09	2.19	2.56	54.68
Total	3.05	3.41	2.91	3.01	2.94	3.11	3.34	3.06	2.95	3.01	2.65	3.13	3.15	2.93	3.21	3.34	3.04	3.23	2.64	3.21	3.02	2.79	3.10	70.22
Gender-wise																								
Male	3.07	3.41	2.89	2.98	2.93	3.12	3.36	3.06	2.92	3.05	2.68	3.11	3.12	2.89	3.19	3.28	3.05	3.22	2.65	3.19	2.97	2.81	3.15	70.08
Female	3.04	3.42	2.94	3.04	2.97	3.07	3.29	3.05	2.97	2.95	2.59	3.16	3.19	2.97	3.20	3.40	3.01	3.26	2.61	3.23	3.07	2.78	3.04	70.22
Over all	3.05	3.41	2.91	3.01	2.94	3.11	3.34	3.06	2.95	3.01	2.65	3.13	3.15	2.93	3.21	3.34	3.04	3.23	2.64	3.21	3.02	2.79	3.10	70.22
Marital Status																								
Married	3.09	3.49	2.97	3.07	3.02	3.15	3.44	3.13	3.04	3.07	2.65	3.16	3.16	3.06	3.30	3.48	3.09	3.34	2.62	3.30	3.04	2.80	3.09	71.51

Un-Married	3.01	3.32	2.87	2.94	2.86	3.08	3.25	2.98	2.86	2.96	2.65	3.12	3.16	2.76	3.10	3.19	2.97	3.15	2.68	3.13	3.02	2.79	3.14	68.97
0ver all	3.05	3.41	2.91	3.01	2.94	3.11	3.34	3.06	2.95	3.01	2.65	3.13	3.15	2.93	3.21	3.34	3.04	3.23	2.64	3.21	3.02	2.79	3.10	70.22
Education																								
no formal education	2.81	3.15	2.74	2.81	2.97	3.17	3.13	3.04	2.91	2.98	2.89	3.22	3.11	2.69	3.10	3.09	3.07	2.92	2.87	2.90	2.99	2.92	3.10	68.55
Upto high school	3.14	3.37	2.95	3.08	3.02	3.14	3.20	3.12	3.02	3.02	2.80	3.19	3.23	3.02	3.21	3.21	3.05	3.21	2.84	3.13	3.03	2.83	3.10	70.88
Bachelor's degree	3.09	3.48	3.05	3.04	3.04	3.22	3.45	3.11	3.04	3.02	2.63	3.15	3.22	2.97	3.20	3.42	2.99	3.32	2.63	3.27	3.08	2.84	3.14	71.38
Master's/ profes-sional degree	2.99	3.41	2.84	2.94	2.81	2.99	3.33	2.95	2.79	3.00	2.47	3.05	3.00	2.82	3.21	3.33	3.03	3.25	2.46	3.22	2.86	2.68	3.01	68.42
others	2.88	3.08	2.66	2.70	2.64	3.04	3.11	3.01	2.94	3.01	2.67	3.17	2.92	2.62	3.00	3.03	2.94	2.96	2.75	3.23	3.05	2.89	2.87	67.12
Over all	3.05	3.41	2.91	3.01	2.94	3.11	3.34	3.06	2.95	3.01	2.65	3.13	3.15	2.93	3.21	3.34	3.04	3.23	2.64	3.21	3.02	2.79	3.10	70.22
Occupation																								
Student	3.06	3.50	2.95	3.06	3.02	3.16	3.42	3.15	3.04	3.06	2.67	3.14	3.16	3.07	3.30	3.51	3.08	3.34	2.63	3.27	3.08	2.78	3.06	71.49
Home Maker	2.92	3.30	2.94	2.97	2.93	3.08	3.27	3.02	2.93	2.95	2.68	3.01	3.22	2.72	3.02	3.26	2.94	3.17	2.74	3.12	3.07	2.79	3.15	69.20
Agri. and Allied Activities	2.93	3.23	2.92	2.82	2.83	3.08	3.15	2.90	2.87	2.80	2.74	3.07	3.05	2.69	3.02	2.99	3.03	2.95	2.66	3.00	3.01	2.70	3.23	67.64
Salaried entrepre-neurs/ Self Business	3.11	3.42	2.91	3.01	2.93	3.12	3.34	3.01	2.91	3.07	2.63	3.17	3.15	2.82	3.19	3.21	3.02	3.22	2.66	3.22	2.95	2.88	3.17	70.08
Others	3.12	3.48	2.91	2.92	2.91	3.21	3.59	3.02	3.05	3.02	2.37	3.34	3.18	3.08	3.17	3.48	3.09	3.40	2.56	3.33	3.00	2.81	3.11	71.09

Overall	3.05	3.41	2.91	3.01	2.94	3.11	3.34	3.06	2.95	3.01	2.65	3.13	3.15	2.93	3.21	3.34	3.04	3.23	2.64	3.21	3.02	2.79	3.10	70.22
Annual Family Income																								
Less than Rs. 1,00,000	3.12	3.47	2.98	3.09	3.05	3.14	3.38	3.13	3.03	3.06	2.81	3.16	3.22	3.07	3.27	3.42	3.11	3.29	2.72	3.24	3.03	2.86	3.10	71.72
Rs. 1,00,000 to less than Rs. 4,00,000	3.04	3.39	2.94	3.00	2.93	3.11	3.28	3.05	2.98	3.01	2.55	3.13	3.14	2.91	3.22	3.32	3.06	3.19	2.63	3.18	3.06	2.79	3.10	69.99
Rs. 4,00,000 to less than Rs. 7,00,000	3.07	3.39	2.86	2.98	2.78	3.12	3.35	3.04	2.83	2.96	2.49	3.16	3.13	2.78	3.17	3.26	2.97	3.21	2.54	3.16	2.97	2.74	3.11	69.04
Rs. 7,00,000 to less than Rs. 10,00,000	2.83	3.78	2.82	3.05	2.63	3.40	3.69	2.78	2.71	3.03	2.32	3.42	3.36	3.17	3.14	3.68	3.04	3.72	2.37	3.64	3.33	2.73	3.23	71.85
Above Rs. 10,00,000	2.83	3.13	2.57	2.60	2.53	2.89	3.59	2.53	2.56	2.87	2.53	2.64	2.65	2.40	2.58	3.06	2.55	3.05	2.22	3.27	2.45	2.36	3.02	62.84
Overall	3.05	3.41	2.91	3.01	2.94	3.11	3.34	3.06	2.95	3.01	2.65	3.13	3.15	2.93	3.21	3.34	3.04	3.23	2.64	3.21	3.02	2.79	3.10	70.22
Residential Area																								
Rural	3.05	3.39	2.97	3.08	3.03	3.18	3.31	3.09	3.04	3.02	2.74	3.19	3.22	2.95	3.19	3.34	3.03	3.22	2.74	3.19	3.06	2.85	3.08	70.94
Urban	3.05	3.43	2.86	2.96	2.86	3.05	3.35	3.03	2.88	3.02	2.58	3.09	3.09	2.93	3.24	3.34	3.07	3.23	2.57	3.21	2.99	2.76	3.10	69.68
Over all	3.05	3.41	2.91	3.01	2.94	3.11	3.34	3.06	2.95	3.01	2.65	3.13	3.15	2.93	3.21	3.34	3.04	3.23	2.64	3.21	3.02	2.79	3.10	70.22

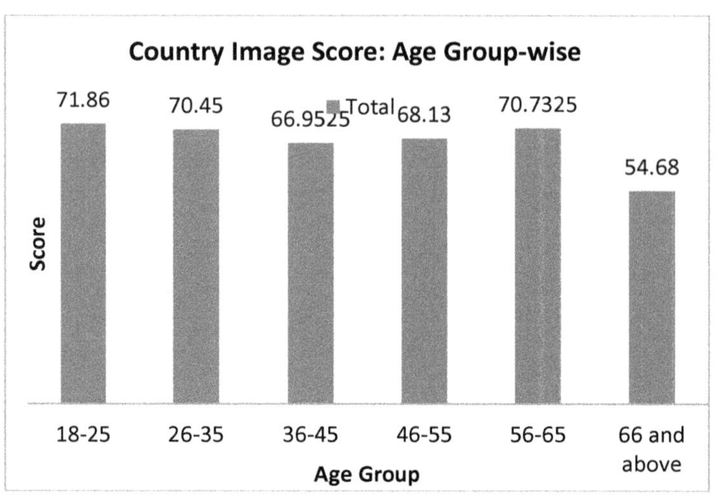
Country Image Score: Age Group-wise
Total
71.86
70.45
66.9525
68.13
70.7325
54.68
Score
18-25
26-35
36-45
46-55
56-65
66 and above
Age Group

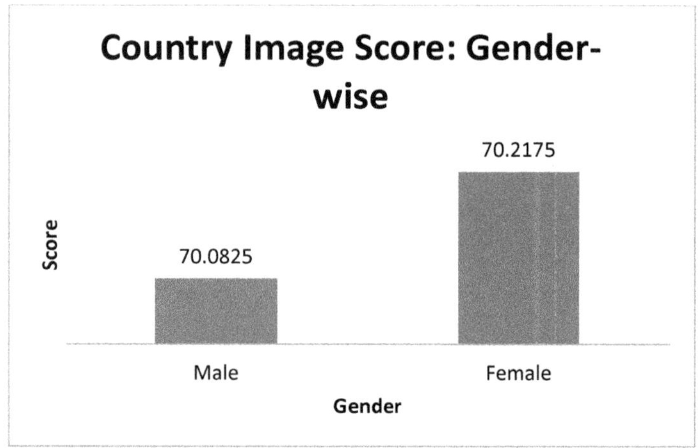
Country Image Score: Gender-wise
70.2175
70.0825
Score
Male
Female
Gender

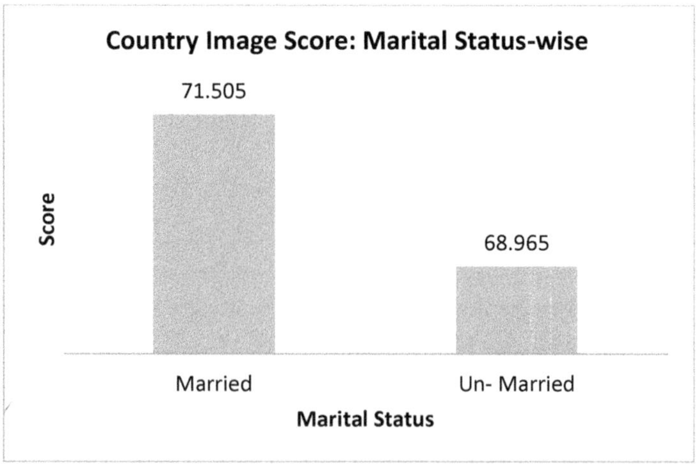
Country Image Score: Marital Status-wise
71.505
68.965
Score
Married
Un- Married
Marital Status

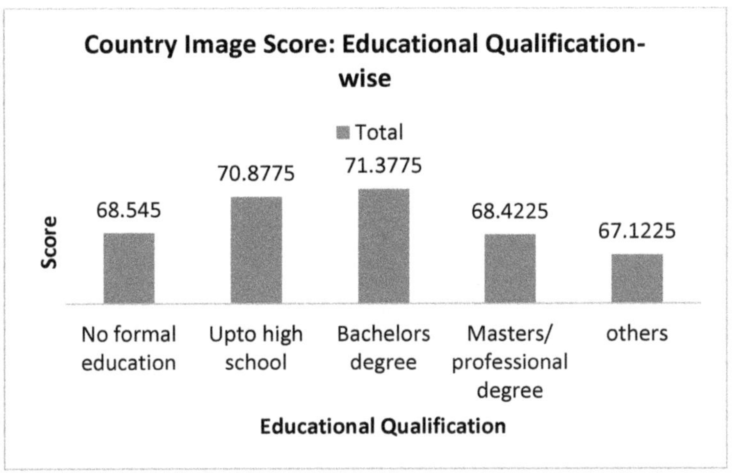
Country Image Score: Educational Qualification-wise
Total
Score
68.545
70.8775
71.3775
68.4225
67.1225
No formal education
Upto high school
Bachelors degree
Masters/ professional degree
others
Educational Qualification

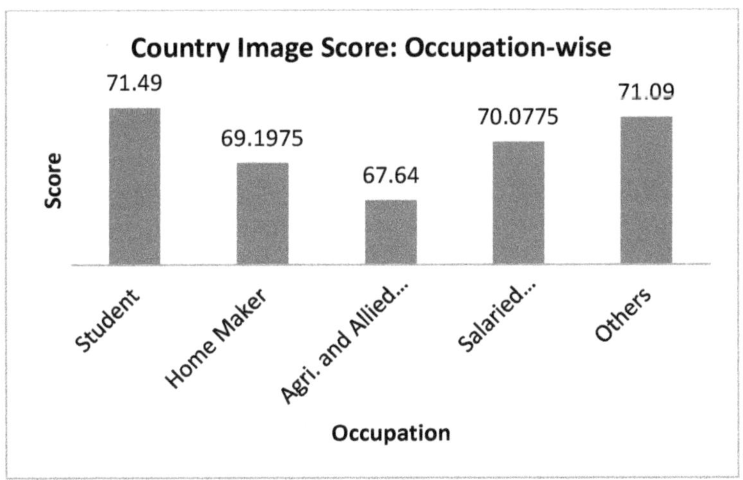
Country Image Score: Occupation-wise
Score
71.49
69.1975
67.64
70.0775
71.09
Student
Home Maker
Agri. and Allied...
Salaried...
Others
Occupation

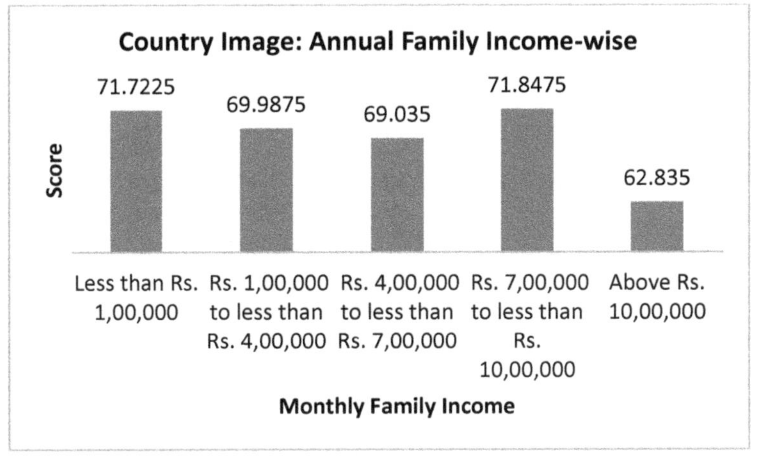
Country Image: Annual Family Income-wise
Score
71.7225
69.9875
69.035
71.8475
62.835
Less than Rs. 1,00,000
Rs. 1,00,000 to less than Rs. 4,00,000
Rs. 4,00,000 to less than Rs. 7,00,000
Rs. 7,00,000 to less than Rs. 10,00,000
Above Rs. 10,00,000
Monthly Family Income

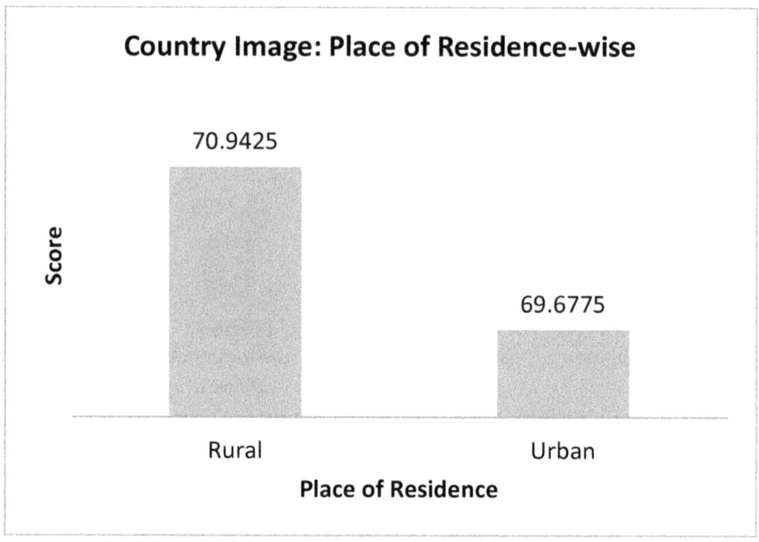

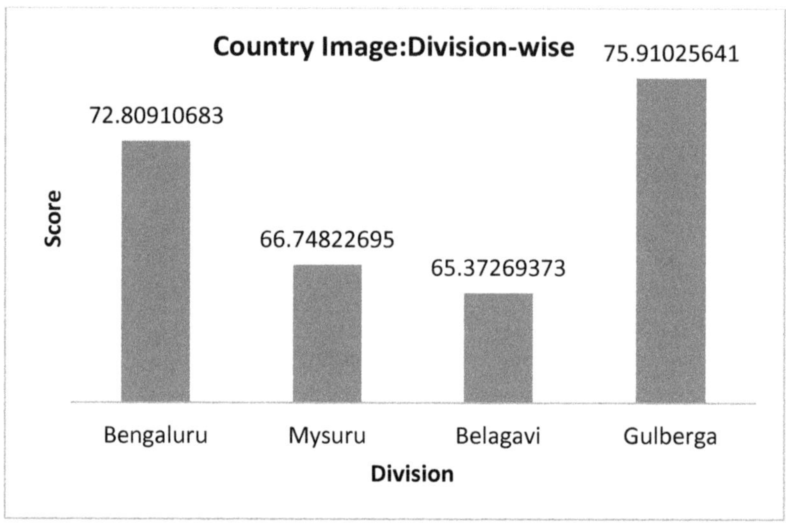

Fig. Average Weighted Score for Country Image as per Demography

The calculated average weighted score for CI: 70.22, which is at the moderate level and there found to be some of the demography variables influencing this CI also as that of CET. The average CI score does not vary significantly as per the age group, gender, marital status and place of residence. However, it is noticed from the figure that the CI varies as per educational qualification, occupation and annual family income. The respondents with no formal education and who are agriculture pursuits have a very low score for CI and their perceive image on India as a country is relatively poor. Similarly, the CI score is found to be lowest among the respondents above Rs.10 lakhs as annual family income.

5

CHAPTER

Findings of the Study

The summary of findings from the study is presented below:

The rigorous literature review revealed that the variables such as awareness level of Indian consumers on coffee/tea producing countries, quality and brands, criteria in evaluation of coffee/tea brand and the relevance of country of origin, attitude towards foreign brands, consumer ethnocentrism and the perceived country image have a greater influence in brand preference and purchase intention.

The review also revealed that there are several studies across the globe on these variables. Unfortunately, these variables are less understood for Indian consumers in general and coffee/tea consumers in particular.

The outcome of the measurement on these variables is depicted in figure 5.1.

Awareness Level of Consumers:

- ➢ The awareness level of consumers on coffee/tea regarding the places in India where they are grown, countries where they are produced, foreign brands and quality of Indian produce do not vary significantly among coffee and tea.
- ➢ The awareness level on the places in India where coffee/tea produced is found to be very poor. Though the overall percentage of respondents who have a complete awareness on the places where coffee and tea grown is found to be very less, the awareness level vary as per the age group. Youngsters relatively have a better awareness and the level does not vary according to the gender and marital status.
- ➢ It is found that the consumers belonging to the higher family income and higher education are better aware of compared to their counterparts. The homemakers have very poor awareness on the places where coffee and tea is grown. Unfortunately, they are the one who make decision related to shopping coffee/tea.

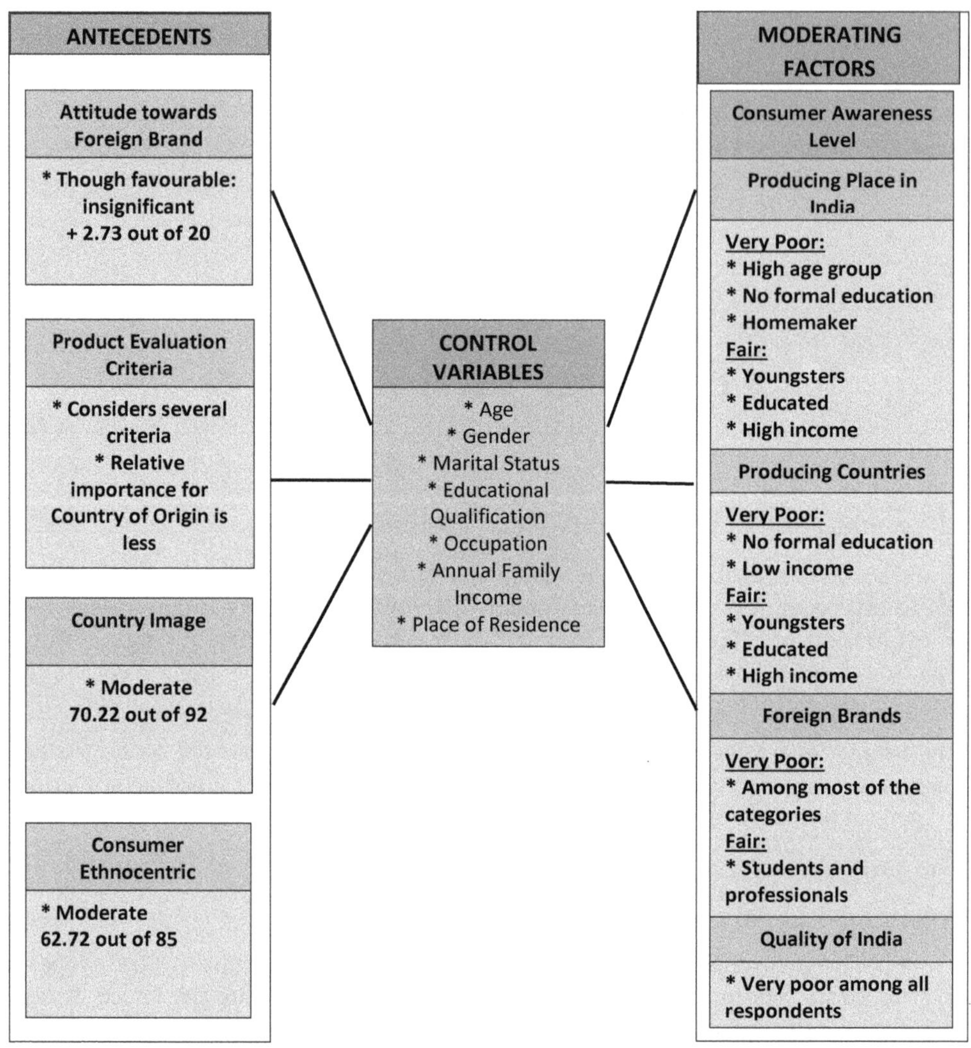

Fig. 5.1: Summary of Findings

- It is found from the study that the consumers' awareness level on the countries where coffee and tea is produced is very less. And the awareness level does not vary much for coffee and tea. The percentage of consumers who are aware of the countries where coffee/tea grown is less than those who are aware of places in India.
- The awareness level on countries also varies as per the age group, educational qualification, and annual family income. Students relatively have a better awareness than consumers belonging to other occupation and consumers with higher family income, the awareness level is also found to be higher.

- The awareness level on the foreign brands of coffee and tea among the consumers in the study area is found to be very less and it varies as per the age group. The lower age group has relatively better awareness than higher age group. It is also found that consumers of higher income group have awareness level fairly.
- Consumers who are aware of the places where coffee /tea grown in India, even they do not have much awareness on the quality of products from these places. The low awareness level of consumers acts as a major barrier for promoting on country of origin labelling.

Importance of Country of Origin Labelling in Brand Evaluation

It is found from the study that while buying coffee/tea, the consumers of the study area consider not just one criterion for brand evaluation. They consider as many as 11 attributes and the relative importance vary among the various demography. The attribute 'taste' is given highest importance and 'country of origin' the least importance among the listed attributes. All the listed criteria were scored in the range of 3 – 4, that is fairly important to very much important.

Attitude Towards Foreign Brands:

- The attitude of consumers in the study area is found to be bit favourable. It is derived from the study that the average score calculated as + 2.73, out of a maximum possible favourable score of +20. The consumers have a slight favourable attitude towards foreign brands of coffee/tea, though the score is very negligible. Most of the respondents had a belief that the use of foreign brands elevates their social status.
- It is found from the study that the consumers attitude varied according to the demography. Consumers of middle age group have higher favourable attitude than their counterparts in the upper group and lower group. It is also found that females and unmarried consumers' attitude is slightly favourable than their counterparts.
- The study revealed that the favourable towards foreign brand comes down with advancement in education and profession. The uneducated and agri & allied agri. groups have favourable attitude than their counterparts.

Consumers' Ethnocentric Tendency

- The Consumers' Ethnocentric Tendency (CET) score for the consumer in the study area is measured as 62.72, out of a maximum of 85, which is at a moderate level. Unlike the inferences from other studies, the study revealed that the CET score is higher among younger age group consumers belonging to 18 – 25 age group.
- The CET score does not change with the gender and the same is lowest among the professional category of consumers and the rich class of consumers with annual family income of more than Rs.10 lakhs.

Consumers' Perceived Country Image

- The score for consumers' perceived Country Image (CI) in the study area is measured as 71.22, out of a maximum of 82, which is at a moderate level.
- Consumers of North Karnataka have a high perceived image of the country and Mysuru with very low CI. Consumers at the age group of 36 – 45 have low CI than upper and lower age group consumers. The married consumers have slightly higher CI as compared to their counterparts.
- Consumers who are not educated much have a poor perception about the country. The CI varies with occupation, professionals and businessmen rating India as a successful country, on the contrary, home makers relatively have low CI.
- The CI does not vary among male and female consumers and urban and rural consumers.

The Current Promotional Practice followed by the Respective Boards:

It is observed that the domestic promotion is done for coffee and tea by Coffee Board and Tea Board respectively. The promotional programmes initiated and implemented by these boards are presented below.

Domestic Promotion by Coffee Board

- To enhance the domestic market consumption, the Coffee Board has schemes under "Market Development" viz. domestic promotion and market research and intelligence.
- The growth in domestic market is facilitated through promoting awareness of coffee and consumption of pure coffee through generic promotion campaigns using mass media, participation in festivals, taking part in exhibitions in different parts of the country, etc.
- For traders/ processors to identify the coffee grown at various regions with different altitude, the Board has come out with logo pertaining to different regions and a common logo for "Coffee India" brand. Figure below shows the logo.

Fig. 5.1 Logo as Trade Mark for Coffee India and Region specific Coffee

➢ The Board also has come out with logo for specialty coffee such as Monsooned Malabar, Mysore Nuggets Extra Bold and Robusta Kaapi Royale. The logo which is being enforced is usually used by the traders of export market and rarely reaches to the consumer.

Domestic Promotion by Tea Board

India is the largest producer and consumer of black tea in the world. India, however, is not the largest per capita consumer. Though tea enjoys over 90% penetration among the country's population, it is generally perceived to be an "addictive" beverage. There is lack of awareness about the positive health benefits of tea. Moreover, it is seen that tea is not considered to be an aspirational beverage among the youth of India. In order to dispel myths about tea and increase its popularity amongst the youth, Tea Board is formulating massive domestic promotional campaigns.

The principle objective of these domestic promotional campaigns is to reinforce the image of tea in the minds of the rural Indian populace, as well as promote tea as the preferred beverage among the upwardly mobile middle class urban youth. To this end, Tea Board has come up with a new mantra, which aptly describes the feeling of well-being one has after sipping a cup of tea:

चाय पियो मस्त जियो !

Chai piyo mast jiyo !

Some of the rationale for undertaking such promotional activities is given below:

➢ Undertaking domestic campaign aimed at increasing per capita tea consumption

➢ Generic campaign in print & electronic media based on wellness benefits & lifestyle aspects

➢ Promotion of single-origin teas in India through road shows, generic as well as specialty tea campaigns, protection and promotion of various intellectual properties of the Board (Darjeeling GI, Assam Orthodox GI, Nilgiri Orthodox GI etc)

➢ Viral advertising through established social networks like Twitter, Facebook for propagating the diversity & richness of Indian tea during domestic promotion;

➢ Additional focus would be on initiating and sustaining promotional activities in the rural markets.

➢ Establishment of Tea Boutiques/Lounges at different locations to promote teas of different origin and to create tea culture habits amongst youth

➢ Focus on the Health benefits of tea to the general public

It was also noted that the board has developed logo for "India Tea" and for region specific teas which acts as a trademark/copy rights. The figure below shows these logos.

Fig. 5.2: Logo as Trade Mark for India Tea and Region specific Tea

The permission to use Assam Tea, Darjeeling Tea, Nilgiri Tea logo is given only to the packers/blenders. The permission is obtained through submitting a formal application to the Tea Board by paying a prescribed fee. The permission thus obtained is valid only for one year and after expiry it has to be renewed.

Fig. 5.3: Logo as Copyrights for Region Specific Tea

The Board issues licenses for use of Assam Orthodox or Nilgiri Orthodox Word and Logo as per the Certification of Trade Mark Regulation. All producers, exporters, blenders, packers, Importer, bulk supplier wholesaler, auctioneer, owner retail store and Tea Boutique are required to obtain license for using the word and/or logo mark of Assam Orthodox/Nilgir Orthodox.

Recommendations

Considering the current scenario it is necessary to initiate relevant policies and strategies to protect the domestic market and the interest of the consumers. The inferences of the study bring in clarity that both coffee and tea can be promoted on the country of origin labelling to reach this desired objective. A three pronged strategy required for this successful promotion and the same is depicted in the figure below:

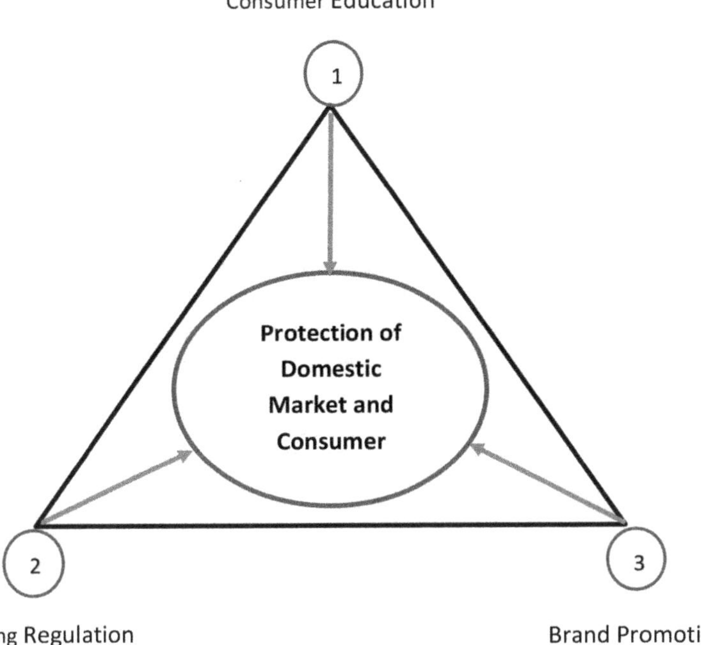

Fig. 6.1: Three pronged strategy for protection of domestic market and consumer.

A. Consumer Education

As the consumers' awareness level on the product origin and the available brands and its quality is very low, it is necessary to create awareness among the consumers so as to empower them with information and rights which enables consumers to take the correct decision.

Awareness has to be created with respect to the places in India where coffee and tea is grown, the special characteristics of coffee / tea grown at different elevations and agro-climatical conditions, soil, traceability, the plant, harvesting method etc.

The consumers also need to be given awareness about preparing coffee/tea, the required intake of coffee/tea with quantity and the frequency of having the same. This helps them to go for a responsible consumption.

Usage of Labels for Consumer Education

When consumers purchase a pack of coffee or tea, they look not just the product, but something more. Consumers want to have connection to their products in their lives. Labels on packs can serve a major role in bringing in more transparent information about the origin, farm and elevation and processing method, etc.

Using visual appeal such as colour, photography, artistic renderings, certification seals, company logos, maps, etc., makes consumer not just identify their brands, but play the role of disseminating information and educating the consumer.

Since the labels act as signaling mechanism by which consumers can get assurances on the quality standards, required awareness may be created so that consumers pay utmost importance to the labelling and demand information regarding the place and how coffee/tea is produced, processed and packed.

Usage of Technology for Interacting with Consumer;

Making provision in the label, inviting consumers for an interaction will be one among the ways suggested to create awareness among consumers. With the integration of digital technology and smart phone, consumers can easily get connected to the regions where coffee/tea is produced and the product characteristics. QR codes – coded arrays of black and white squares read by smart phone as URLs, makes consumers to initiate interaction pre and post purchase. This is one among the suggested methods for educating the young and techno savvy consumers.

Consumer Education for Changing Evaluation Pattern:

Consumers must be educated in a way to change their brand evaluation pattern. Necessary awareness must be made to give priority to country of origin' which evaluating the brands for coffee / tea. As of now, taste is given high priority. The education in this way help them to be carried away by the taste by compromising on the quality and safe consumption.

Engaging Voluntary Groups:

There are several bodies such as consumer organization, voluntary groups, etc. playing a significant role in consumer education in India. They have the means to provide consumers with information and help them to make informed choices.

For instance, there are several NGOs and SHGs in the coffee growing area engaged in organizing awareness programme and promoting coffee. Rajarajeshwari Self Help Group in Thalacauvery and Women Coffee Promotion Council (WCPC) in Sakleshpur, Coorg, Karnataka are few such. They play a pivotal role in coffee promotion activities making coffee as a popular beverage. The promotional activities include:

- Creating coffee awareness through organizing seminars.
- Popularizing the benefit of drinking coffee.
- Demonstrating the art of coffee making.
- Spreading the culture of coffee to non-traditional areas.
- Promotion through placards, banners, slogans, bill boards, etc.

There are similar groups found in tea growing area also. The consumer education through a proper agency will therefore enable to protect the consumers' interest and also the domestic market.

B. Regulation on Mandatory Labelling

The findings from the present study provide insight into consumers' interest in understanding of, and expectations for origin labelling of coffee/tea. Making origin labelling as mandatory for coffee/tea produced, processed, packed and marketed within the country would be the second suggested strategy through which the consumers' interest and domestic market can be protected.

As per Food Laws every packaged food article has to be labeled and it has to be labeled in accordance to the law applicable in the country of the user. Every packaged food article for the domestic use has to be labeled in accordance to the related Indian Food Law i.e. Food Safety and Standards (Packaging and Labelling) Regulations, 2011, notified by Food Safety and Standards Authority of India (FSSAI).

In order to safe guard the interest of the consumer, The Food Safety and Standards (Packaging and Labelling) Regulations, 2011, provides that every packaged food article has to be labeled and it shall provide the following information –

- The name of Food
- List of Ingredients,
- Nutritional Information,
- Declaration regarding Veg or non-veg,

- Declaration regarding Food Additives,
- Name and complete address of the manufacturer or packer
- Net Quantity,
- Code No,/Lot No./Batch No.,
- Date of manufacture or packing,
- Best Before and Use By Date,
- Country of Origin for imported food and
- Instructions for use

In addition to the above information the manufacturer or the packer has to also ensure that the label complies with the general requirements of labelling prescribed under the regulations I.e. the label should not become separated from the container, contents on the label shall be correct, clear and readily legible and shall be in English or Hindi language, etc.

However, there is no specific mentioning on providing information on "country of origin for products (coffee/tea) produced within the country". Both Coffee Board and Tea Board have developed a logo that depicts Brand India Coffee and Tea respectively. These logos may be in the same form or a new improved one may be made mandatory packaging label to be followed by all processing and marketing companies using Indian Coffee/Tea.

It is crucial for consumers that the withdrawal of unsafe products from the market, or the recall of products that hold potential risks to health and safety, is done as quickly as possible. Labelling done as mandatory and the usage of proper technology will help consumers to trace the origin and ensure food safety.

Making the logo and origin labelling as mandatory in the domestic market helps in protecting the identity till coffee/tea reaches the consumer. Granting the license and permission to the coffee/tea marketers who meet the desired requirement, not just help them to promote their brand, it also puts pressure on the planters, processors and marketers to strictly adhere to the standards set. Thus, it ensures that the consumer is not exploited with inferior quality products.

C. Branding and Brand Promotion

Since labelling acts as a vehicle for marketing, suitable promotional tools may be deployed to leverage the commercial benefits of the same. Promoting coffee/tea on country of origin labelling helps in creating a strong distinction in the market place and acts as a strong barrier for the late entrants into this competitive market. Therefore, it is suggested that policy measures may be initiated to build a brand 'India Coffee' and 'India Tea'. Generic promotion should also be made to increase the domestic consumption of tea and coffee. To make the branding on 'country of origin labelling' effective, generic promotion should also be made across the country to build the image of the country. Following are some of the specific suggestions:

Holistic Approach in Branding

Considering the long value chain in coffee/tea and the value contributed by different market functionaries, the branding efforts is to be done with a holistic approach. Holistic branding recognizes all the participants right from planter, through processor and to the marketer and their activities. For a successful branding based on origin, it is important that all stakeholders in the value chain actively share the responsibility and promote the same.

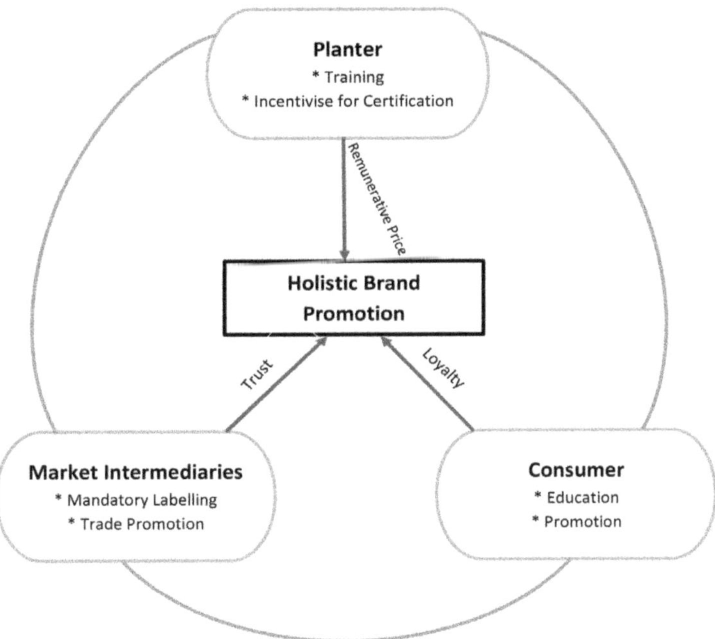

Fig 6.2: Holistic approach in brand promotion

The planters play an important role in making the brand promotion on 'origin' possible. The coffee/tea planters must strictly follow the standard practices in farming, harvesting and value addition, if they do. This ensures that there is a standardization of the produce at their level, which is the basic requirement for branding effort. The effort put and the time invested by the planters, more specifically small holders would help them to improve their bargaining power and build competitiveness in the market place. Moreover, the branding effort would raise the value and image of coffee/tea and fetches remunerative price which would benefit the planters. It is suggested to incentivize the individual planters or collectives of small growers of coffee and tea, who show willingness in maintaining the consistency in quality through any appropriate certification.

Creating a logo and the vocabulary and making it mandatory to the processors and marketers would bring in more trust, confidence and comfort among the intermediaries. Therefore, promoting the brand among the market intermediaries will remove the apprehensions of these intermediaries and makes them to actively participate in trading and marketing.

Brand promotion may also be aiming at the final consumer to establish brand image and loyalty among the consumers. This will create consumer demand and also acts as a strong barrier for any other inferior brands of coffee/tea entering into the Indian market.

Generic Promotion

Suitable promotional strategies must be initiated to bring in awareness among the consumers on the importance of 'origin' among the consumers.

As far as coffee is concerned, domestic market is a growing one and globally, the growth is shifting from consuming country to producing country. It is true for India also. Promotional efforts are to be made to increase the per capita consumption. Under this circumstance, it is the basic responsibility of the policy makers to guide and educate the consumers to indulge in responsible consumption. The contribution made by the Board through participating in exhibitions, organizing coffee festivals and campaigns, etc are very much appreciable. However, more promotional efforts are to be made to build brand 'India Coffee'.

The scenario for tea is very much different from that of coffee. The domestic consumption for tea increases more specifically owing to the growth in the Indian population. Though tea is consumed by both rich and poor, the per capita consumption remains to be on the lower side compared globally. Therefore, it is suggested to have more promotional measures to enhance the image of tea, importance of tea consumption through product development and branding.

Promoting Country Image

It is not just enough that the planter maintains consistency in quality of coffee/ tea and/or the processor and marketer build the corporate image and his brand image for a successful promotion of coffee/tea on country of origin labelling. It is very much essential to create a positive perception about India as a country among the consumers. Therefore, generic promotion to bring the image of the country among the citizens of India is imperative. Branding process may be done by aiming at strengthening the existing image or creating a new and different one. It is suggested that the existing tag line 'Incredible India' may be strengthened more through mass campaigning and the image of the country may be strengthened. Efforts must be made to strengthen the people's confidence on political system, stabilize the economy, technological advancement, etc. which can help in build the image of the country, not only within also globally.

7

CHAPTER

Conclusion

The study has brought in new insights in understanding the consumers' brand evaluation pattern and their behavior. Consumers in general found to be having more affinity towards products made in India. There is a positive attitude towards domestic brands of coffee/tea. The consumer ethnocentric tendency and the perceived country image favours the promotion of coffee/tea on country of origin labelling.

The study will be of great relevance to planters and marketers in maintaining quality and share branding responsibility marketers to initiate promotional programmes on COOL and policy makers to plan and implement appropriate policies to regulate the market and educate the consumer. Most important of all, the consumers are the major beneficiaries of the outcome of the study.

By creating the desired of level of awareness on the places of coffee/tea grown, brands available and the importance to be given to the 'country of origin labelling', we can develop a positive attitude among the consumers towards domestic brands of coffee/tea. Promotion of coffee/tea on 'country of origin labelling' through consumer education, mandatory regulation and holistic brand promotion approach will help in protection of the consumer and the domestic market in this globalized economy.

There is an increasing concern in strengthening market competitiveness for coffee and tea in the international market. The brand promotion efforts for domestic market acts as a base for building brand image in the international market on 'country of origin labelling' which is very much the need of the hour.

However, there are few limitations of the study which poses a serious barrier on generalizing the inferences. The study is done restricting to coffee and tea only and therefore, the findings cannot be generalized to other products. It would therefore be particularly enlightening for future research to replicate this study to other product categories such as durables, FMCGs and services and the results may be compared.

The study has been carried out across the different class of consumers in Karnataka state, However, socio-economic profile of consumers and the consumption pattern for coffee and tea differs largely across the different states of India, results a serious limitation on the relevance of the inferences to other places. Also, the consumption pattern for coffee differs largely from that of tea. Since the study covered both coffee and tea, more specific questions on coffee/tea consumption could not be asked. Therefore, future in-depth studies may be carried out across the country, exclusively for coffee and tea separately.

Another limitation of the present study is the framework of study, where the research variables, such as awareness level, consumers' evaluation criteria, attitude towards foreign brands, consumer ethnocentrism and country image were measured independently and what is the extent of influence of these variables on purchase intention was not measured. More in-depth study is required in this direction by incorporating additional variables an with an alternate conceptualization of the research model.

More importantly, such studies on promotion based on 'country of origin labelling' found to be very much relevant to international marketing strategy. Though the inferences and the outcome of the study acts as a base for international promotion, the same cannot be replicated for export market of coffee and tea. Therefore, an in-depth study on the feasibility of branding on 'country of origin labelling' need to be carried out separately for coffee and tea among the consumers of coffee and tea in the emerging importing countries.

Annexure

Study on Promoting Indian Coffee and Tea on Country of Origin Labelling

This survey has been carried out as a part of assessing the feasibility of promoting Indian Coffee and Tea in the domestic market using the country of origin labelling. You are requested to give your free opinion to the following questions. The responses will be kept confidential and will be used only for the intended purpose.

Thank you for participating in our survey. Your feedback is important.

Questionnaire

District: ___________ City/Town/Taluk :___________

Part A: General Awareness level on Coffee/ Tea-

1. Are you aware of the places in India where tea is grown?

   a) No awareness b) Little awareness c) Complete awareness

2. Are you aware of the places in India where coffee is grown?

   a) No awareness b) Little awareness c) Complete awareness

3. Are you aware of the countries where tea is grown?

   a) No awareness b) Little awareness c) Complete awareness

4. Are you aware of the countries where coffee is grown?

   a) No awareness b) Little awareness c) Complete awareness

5. Are you aware of foreign brands of coffee/ tea?

   a) No awareness b) Little awareness c) Complete awareness

6. Are you aware of the quality of the Indian tea?

   a) No awareness b) Little awareness c) Complete awareness

7. Are you aware of the quality of Indian coffee?

   a) No awareness b) Little awareness c) Complete awareness

Part B: Evaluation Criteria while buying Coffee/ Tea -

8. Following is the criteria generally used while purchasing Coffee/ Tea. Kindly rate these criteria Please rate the following on a scale of 1- 5. (1 being the lowest importance and 5 being the highest.)

Attribute	1	2	3	4	5
Aroma					
Availability					
Brand					
Country of origin					
Cuppage					
Ease of use (instant coffee, tea bags. Etc.)					
Flavour					
Information label (Packaging date, nutritional info, etc.)					
Packaging					
Price					
Taste					

Part C Attitude towards foreign brands of Coffee/ Tea.

9. Following are the statements pertaining to your belief on foreign brands of Coffee/ Tea.

Please rate the appropriate option (on a scale of -2 to 2.) (-2: strongly disagree,-1: disagree, 0: neutral, 1: agree, 2: strongly agree)

Sl. No.	**Statement**	**-2**	**-1**	**0**	**1**	**2**
1	Foreign brands of Coffee/ Tea are of superior quality.					
2	Price of foreign brand Coffee/ Tea is relatively higher.					
3	Consumption of foreign brand Coffee/ Tea enhances my social status.					
4	Foreign brand Coffee/ Tea are available in more varieties and flavours.					
5	Foreign brand Coffee/ Tea come with a lot of promotional schemes.					

Sl. No.	Statement	-2	-1	0	1	2
1	I give importance to quality while buying Coffee/ Tea.					
2	I give importance to price while purchasing Coffee/ Tea					
3	I give importance to social status while consuming Coffee/ Tea.					
4	I give importance to variety of Coffee/ Tea.					
5	I give importance to promotional schemes while buying foreign brand of Coffee/ Tea.					

Part D : The Consumer Ethnocentrism

10. Rate the following on a scale of 1 to 5. (1 being the lowest and 5 being the highest.)

Attribute	1	2	3	4	5
Indian people should always buy Indian tea/ coffee instead of imported tea/ coffee.					
Only those products that are unavailable in India should be imported.					
Buy Indian- made products, keep India working.					
Indian products, first, last, and foremost.					
Purchasing foreign made tea/ coffee is un-Indian					
It is not right to purchase foreign tea/ coffee because it puts Indian out of job					
A real Indian should always buy India-made tea/ coffee					
We should purchase tea/ coffee manufactured in India instead of letting other countries get rich off us					
It is always best to purchase Indian tea/ coffee					
There should be very little trading or purchasing of tea/ coffee from other countries unless out of necessity					
Indian should not buy foreign tea/ coffee because it hurts Indian business and causes unemployment					
Curbs should be put on all imports					
It may cost me in the long run but I prefer to support Indian tea/ coffee					
Foreigners should not be allowed to put their tea/ coffee on your market					
Foreign tea/ coffee should be taxed heavily to reduce their entry into India					
We should obtain from foreign countries only those products that we cannot obtain in our country.					
Indian consumers who purchase products made in other countries are responsible for putting their fellow Indians out of work.					

Part E: Country Image

11. Please find description of the country listed below. Select your closest opinion on a 4 point scale (1: not at all to 4 : Very much.

Sl.No	India is a	1	2	3	4
1	Successful country				
2	Country with a lot of talent				
3	Country of fair and honest people				
4	Educated, civilized country				
5	Country of social justice				
6	Country of human freedom				
7	Country with a glorious and rich history				
8	Much-suffered country				
9	Country of law and order				
10	Country of entrepreneurs				
11	Depressed, pessimistic country				
12	Hard-working country				
13	Country with solidarity				
14	Country with great sports achievements				
15	Country with great scientific achievements				
16	Country with great culture				
17	Country with great economic achievements				
18	Democratic country				
19	Decent, clean country				
20	Country rich in beautiful landscapes				
21	Cheerful country				
22	Developed country				
23	Fast-developing country				

Part F: Demography Profile of Respondents

12. Age

a) 18- 25 b) 26- 35 c) 36- 45

d) 46- 55 e) 56- 65 f) 65 and above

13. Gender

a) Male b) Female

14. Marital Status

    a) Single b) Married

15. Education

    a) No formal education b) Up to High School

    c) Bachelor's degree d) Master's / Professional degree

    e) Others__________________________

16. Occupation

    a) Student b) Home maker

    c) Agri. And allied activities

    d) Salaried Entrepreneurs/ Self Business

    e) Others_________________

17. Annual family Income

    a) Less than Rs. 1,00,000 b) 1,00,000, less than 4,00,000

    c) 4,00,000, less than 7,00,000 d) 7,00,000, less than 10,00,000

    e) Above 10,00,000

18. Place of residence-

    a) Rural area b) Urban area

19. Name (optional) : Mob. No. (Optional):

20. E mail Id (optional) :

References

Abbott R., 1997 Food and Nutrition Information: A Study of Sources, Uses, and Understanding. British Food Journal, 99(2): 42-44

Aaker, D.A. (1991). Managing Brand Equity: Capitalizing on the Value of a Brand Name. The Free Press, New York, NY.

Agarwal, S. and Sikri, S. (1996) 'Country image: consumer evaluation of product category extensions', International Marketing Review 13(4): 23–39.

Agrawal (1994): Agrawal, Sunitha: Marketing Research, Delhi. Global Business Press

Allred, A. – Chakraborty, G. – Miller, S. J. (1999): Measuring Images of Developing Countries: A Scale Development Study. Journal of Euro-marketing. Vol. 8. No. 3. pp. 29-49.

Avraham, E. – Ketter, E. (2006): Media Strategies for Improving National Images during Tourism Crises, in: Kozak,

Bannister, J. P. & Saunders, J. A. (1978). UK consumers' attitudes towards imports: The measurement of national stereotype image, European Journal of Marketing, 12 (8), 562 – 570.

Batra, Rajeev, Venkatram Ramaswamy, Dana L. Alden, JanBenedict E.M. Steenkamp, and S. Ramachander (2000), "Effects of Brand Local and Nonlocal Origin on Consumer Attitudes in Developing Countries," Journal of Consumer Psychology, 9 (2), 83–95.

Brijs, K. – Bloemer, J. – Kasper, H. (2011): Country-Image Discourse Model: Unraveling

Meaning, Structure, and Function of Country Images. Journal of Business Research, Vol. 64 pp. 1259-1269.

Campos, S., Doxey, J., & Hammond, D. (2011). Nutrition labels on pre-packaged foods. A systematic review. *Public Health Nutrition, 14*(8), 1496-1506.

Caswell JA, Padberg DI (1992). Toward a more comprehensive theory of food labels. Am. J. Agric. Econ. 74(2):460-468.

Chang, H.H., & Y.M. Liu, 2009. The impact of brand equity on brand preference and purchase intention in the service industries. The Service Industries Journal, 29(12): 1687-1706.

Chen, C., & Y. Chang, 2008. Airline brand equity, brand preference, and purchase intentions-The moderating effects of switching costs. Journal of Air Transport Management, 14: 40-42.

Chon, K.S., (1991), "Tourism Destination Image Modification Prosess BMarketing Implications", *Tourism Management*, March 1991, pp. 68-72

Chung, J.E., D.T. Pysarchik, & S.J. Hwang, 2009. Effects of Country-of-Manufacture and brand Image on Korean Costumers' Purchase Intention. Journal of Global Marketing, 22: 21-41.

Cowburn G, Stockley L (2005). Consumer understanding and use of nutrition labelling: a systematic review. Public Health Nutr. 8(01):21-28.

D'Astous, A. & Boujbel, L. (2007). Positioning Countries on Personality Dimensions: Scale Development and Implications for Country Marketing. *Journal of Business Research*, 60(3), 231-239.

Desborde, R. D. (1990): Development and Testing of a Psychometric Scale to Measure Country-of-Origin Image. Florida State University (UMI), Michigan.

De Mooij (2004): Consumer Behaviour and Culture: Consequences for Global Marketing and Advertising. Thousand Oaks, CA: Sage Publications.

Good & Huddleston (1995): Good, L. K., & Huddleston, P. (1995). Ethnocentrism of polish and russian consumers: Are feelings and intentions related?. *International Marketing Review,* 12(5), 35-48. http://dx.doi.org/10.1108/02651339510103047.

Graby, F. (1993): Countries as Corporate Entities in International Markets, in Product – Country Images. Impact and Role in International Marketing,

Grunert, K. G., Wills, J. M. & Fernández-Celemín, L. (2010) Nutrition knowledge, and use and understanding of nutrition information on food labels among consumers in the UK. *Appetite*, 55, 177-189.

Gunn (1972): Gunn, C.A., "Vacationscape: Designing Tourist regions" Austin: University of Texas Press, 1972

Guthrie, J.F., Fox, J.J., Cleveland, L.E., Welsh, S., 1995. Who uses nutrition labeling, and what effects does label use have on diet quality? *Journal of Nutrition Education*, 27 (4), 163–172

Han, C.M. (1989). Country image: Halo or summary construct? Journal of Marketing Research, May, 26: 222-29.

Hieke, S., & Taylor, C. R. (2012). A critical review of the literature on nutritional labeling. *Journal of Consumer Affairs, 46*(1), 120-156.

Huddleston, Linda & Lesli (2001): Consumer ethnocentrism, product and polish consumers' perceptions of quality. International Journal of Retail and Distribution Management, 2995), 236-246. http://dx.doi.org/10.1108/09590550110390896.

Jaffe, Eugene D. – Nebenzahl, Israel D. 2001: National Image and Competitive Advantage – The Theory and Practice of Country-of-Origin Effect. Copenhagen Business School Press,

Jaffe, E.D. and Nebenzahl, I.D. (2006) *National Image and Competitive Advantage: The theory and Practice of Place Branding.* Copenhagen: Copenhagen Business School Press

Jenes, B (2007): Connection between the ecologically oriented consumer behaviour and country image. Marketing és Menedzsment, 2007/ 6. pp. 34-43.

Joint FAOWHOCAC (2009). Food labelling: Food & Agriculture Org. www.fao.org.

Juric *et al.*, (1995): "Open market aftershocks: Czech and Slovak attitude towards brands". Towards A market economy: Beyond the Point of No return, Second East and Central European Conference, ESOMAR, Warsaw, April 1995

Katharina Petra Roth (2006): The Impact of Consumer Ethnocentrism, Consumer Cosmopolitanism and National Identity on Country Image, Product Image and Consumers' Purchase Intentions, dissertation EMAC Doctoral Colloquium, Athens.

Kaynak & Cavusgil (1983): Consumer attitudes towards products of foreign origin: do they vary across product classes? International Journal of advertising, 2,147-157.

Kaynak E & Kara, A. (2002). Consumer's perceptions of foreign products. European Journal of Marketing, 36(7/8), 928-949. http://dx.doi.org/10.1108/03090560210430881

Kaynak, E., Kucukemiroglu, O., & Hyder, A. (2000). Consumers' country-of-origin COO perceptions of imported products in a homogeneous less-developed country. *European Journal of Marketing*, 34(9, 10), 1221–1241.

Keller, K.L. (1993), "Conceptualizing, Measuring and Managing Customer-Based Brand Equity", *Journal of Marketing,* Vol. 57, pp. 1-22.

Kim, C.K. and Chung, J.Y. (1997) 'Brand popularity, country image and market share: an empirical study', Journal of International Business Studies 28(2): 361–387.

Kim S.Y., Nayga R.M. Jr., and Capps O. Jr., 2001. Food Label Use, Self-Selectivity, and Diet Quality. Journal of Consumer Affairs, 35(2): 346-348

Knight, G. A., & Calantone, R. J. (2000). A flexible model of consumer country-of-origin perceptions. *International Marketing Review*, 17(2), 127-145

Kotler, P. – Haider, D. – Rein, I. (1993): Marketing Places: Attracting Investment and Tourism to Cities, States and Nations. Free Press, 1993

Kotler, Haider, & Rein, (1993): Kotler, P., Haider, D. H., Rein, I., "Marketing Places", *THE FREE PRESS*, New York, 1993

Kucukemiroglu, O. (1999). Market Segmentation by Using Consumer Lifestyle Dimensions and Ethnocentrism. An Empirical Study. *European Journal of Marketing*, 335(6), 470–487.

Lantz & Loeb (1996): Country of Origin and ethnocentrism: an analysis of Canadian and American preferences using social identity Theory. Advances in Consumer Research, 23, 374-378. Retrieved from http://ehis.ebscohost.com/

Lee, W., Hong,& Lee, S (2003): Communicating with American consumers in the post 9/11 climate. An empirical investigation of consumer ethnocentrism in the United States. International Journal of Advertising, 22, 487-510. Retreived from http://ehis.ebscohost.com/

Lee & Simon (2006): Cultural Variation in Country of Origin Efects. Journal of Marketing Research, vol. 37 (3), pp. 309-317

Lutz, S., Talavera, O., & Park, S. M. (2008). Effects of Foreign Presence in a Transition Economy, Regional and Industrywide Investments and Firm-Level Exports in Ukrainian Manufacturing. *Emerging Markets Finance and Trade*, 44(5), 82 – 98.

Mc Cullough J., and Best, R., 1980. Consumer Preferences for Food label Information: A Basis for Segmentation. *Journal of Consumer Affairs*, 14(1): 180-182

Mackison, D., Wrieden, W. L., & Anderson, A. S. (2010). Validity and reliability testing of a short questionnaire developed to assess consumers' use, understanding and perception of food labels. *European Journal of Clinical Nutrition, 64*(2), 210-217

Martin, I. M., & Eroglu, S. (1993). Measuring a Multi-Dimensional Construct: Country Image. *Journal of Business Research*, 28(3), 191-210 and foreign products. Ph.D. thesis BUESPA

Mooij & Hofstede (2002): The Roles of Consumer Ethocentricity and Attitude toward a Foreign Culture in Processing Foreign-of-Origin Advertisements. *Advances in Consumer Research*, 23, 436–439

Motameni R. & Shahrokhi M. (1998), Brand equity valuation: a global perspective, Journal of Product and Brand Management, 7, 4, 275-290

Netemeyer., Richard, G., Durvasula, S., Lichtenstein,, & Donald, R. (1991). A cross national assessment of the reliability and validity of the CETSCALE. Journal of Marketing Research, 28(3), 320 - 327. Retrieved from http://ehis.ebscohost.com/

O'Shaughnessy, J., & O'Shaughnessy, N. (2000). Treating the nation as a brand: some neglected issues. *Journal of Macromarketing*, 20 (1), 56-64.

Papadopoulos, N., Heslop, L.A., "Product-Country Images-Impact and Role in International Marketing", *International Business Press,* New York, 1993

Papadopoulos, N., Heslop, L. A., & IKON Research Group (2000). A Cross-national and Longitudinal Study of Product-Country Images with a Focus on the U.S. and Japan. *Working Paper, Report No. 00-106,* Marketing Science Institute: Cambridge.

Petrovici, D. A., & Ritson, C. (2006). Factors influencing consumer dietary health preventative behaviours. *BMC Public Health, 6,* 222

Ravi P. *et al.* (2007). Country image and consumer-based brand equity: relationships and implications for international marketing. Journal of International Business Studies 38, 726–745

Roopam Singh and Ranja Sengupta (2009): The EU India FTA in Agriculture and Likely Impact on Indian Women, http://in.boell.org/sites/default/files/downloads.

Roth M.S. (1992), Depth versus Breadth Strategy for Global Brand Management, Journal of Advertising, 21, 2, June.

Roth M.S. (1995), The Effect of Culture and Socioeconomics on the Performance of Global Brand Image Strategies, Journal of Marketing Research, 32, May, pp.163-175.

Roth, M.S. & Romeo, J.B. (1992). Matching product category and country image perceptions: a framework for managing country-of-origin effects. Journal of International Business Studies, Vol. 23 No. 3, pp. 477-97.

Schooler, R. D. (1965). Product bias in Central American common market. *Journal of Marketing Research,* 2(4), 394-397

Shankarmahesh M.N (2006). Consumer ethnocentrism: an integrative review of its antecedents and consequences. International Marketing Review, 23(2), 146-172. http://dx.doi.org/10.1108/02651330610660065

Shimp, T & Sharma, S.(1987). Consumer ethnocentrism: construction and validation of the cetscale. Journal of Marketing Research, 14, 280-289. Retrieved from http://jstor.org

Shimp, Sharma and Shin (1995): Consumer Ethnocentrism: A test of Antecedents and moderators," Journal of the Academy of Marketing sciences,23 (1), 26-37

Shimp & Sharma (1987): Shimp, T., & Sharma, S. (1987). Consumer ethnocentrism: construction and validation of the cetscale. *Journal of Marketing Research,* 14, 280 - 289. Retrieved from http://jstor.org

Souza-Monteiro DM, Caswell JA (2004). The economics of implementing traceability in beef supply chains: Trends in major producing and trading countries University of Massachusetts, Amherst Working Paper No. 2004-06.

Summer (1996): Summer: Sumner, W. G. (1906). *Folkways: The sociological importance of usages, manners, customs, mores, and morals.* NY: Harper & Row.

Szeles, P. (1998): "The Credit of Reputation. Image and Identity." Star PR

Szeles, P. (1998): A hírnév ereje. Image és arculat. Star PR Ügynökség, Budapest, pp. 81, 93,94, 124, 138.

Thakor, M.V., & Katsanis, L.P. (1997). A model of brand and country effects on quality dimensions: Issues and implications. Journal of International Consumer Marketing, 9(3), 79-100.

Turney, Michael.Images can be natural or constructed perceptions, On - line Readings in PR (http://www.nku.edu/~turney/prclass/readings/image.html) 2000. Ügynökség, Budapest, 1998, pp. 81, 93, 94, 124, 138

Verlegh, P. W. J., & Steenkamp, J.-B. E. M. (1999). A Review and Meta-Analysis of Country-of-Origin Research. *Journal of Economic Psychology*, 20(5), 521-546

Vida and Fairhust (1999): Factors underlying the phenomenon of Consumer ethnocentricity.Evidence for four central european countries. " The international review of Retail distribution and consumer research, 90(4),321-337

Wall,& Heslop (1991): Impact of Country of origin Cues on Consumer Judgements in Multi-Cue Situations: a covariance analysis. *Journal of Academy of Marketing Science.* 19. 105-113.

Wandel M., 1997. Food Labeling from a Consumer Perspective. *British Food Journal*, 99(6): 212-214.

Wang X.H. & Yang Z.L. (2008). Does country-of-origin matter in the relationship between brand personality and purchase intention in emerging economies: Evidence from China's auto industry. International Marketing Review, 25(4), 458-474.

Wozniak S (2010). Has Country of Origin Labeling Influenced Salmon Consumption? Selected Paper prepared for presentation at the Southern Agricultural Economics Association Annual Meeting, Orlando, FL, February 6-9

Yoo, B. and Donthu, N. (2001) 'Developing and validating a multidimensional consumer-based brand equity scale', Journal of Business Research 52(1): 1–14.

Zeugner-Roth, K.P., Diamantopoulos, A., Montesinos. M.A. (2008) Home Country Image, Country Brand Equity and Consumers® Product Preferences. *Management International Review.* 48(5). 578-602.

Jane Shi, —Starbucks, Café Coffee Day Favorites to Conquer India's Emerging Coffee Market,‖ www.2point6billion.com, April 6, 2011

Suparna Goswamy Bhattacharya, Starbucks Could Cause A Veritable Tempest In Local Coffee Mart, http://www.dnaindia.com/money/report, April 11, 2011

http://moneyslash.com, July 10, 2011

Capturing India's Percolating Coffee Market, http://knowledge.wharton.upenn.edu/india/ June 02, 2011

(Footnotes)

1 Bannsiter- Saunders (1978): UK'S consumer attitudes towards imports. The measurement of national stereotype image.European Journal of marketing. Vol 12, No8,pp 562-570.

2 Desborde, R.D (1990); Development and testing of psychometric scale to measure country of origin image. Florida state university.UMI, Michigan.

3 Martin, I.M- Eroglu,S (1993); Measuring a multi-dimensional construct; Country image. Journal of business research, Vol.28,pp. 191-210.

4 Kotler, P- Haider, D- Rein, I (1993): Marketing places: Attracting investment and tourism to cities, states and nations, Free press, 1993.

5 Szeles, P (1998): A Hirnev ereje. Image es arculat. Star PR Ugynokseg, Budaspet,pp 81,93,94,124,138.

6 Allred A- Chakraborthy, G- Miller S J (1999): Measuring images of developing countries; A scale development study, Journal of euro marketing, Vol.8, No3, pp 29-49.

7 Verlegh- Steenkamp (1999): A review and meta analysis of country of origin research, Journal of economic psychology. Vol 20/1999. Pp.521-546.

8 Avraham- Ketter (2006); Media strategies for improving national images during tourism cities in Kozal, M- Andreu, L ed (2006); Progress in tourism marketing, Elsevier, UK,pp.115-128.

9 Brijis, K- Bloemer, J- Kasper,(2011): Country of image discourse model: Unraveling meaning, structure and function of country image. Journal of business research, Vol 64,pp,1259-1269.

www.ingramcontent.com/pod-product-compliance
Ingram Content Group UK Ltd.
Pitfield, Milton Keynes, MK11 3LW, UK
UKHW021953270726
14060UKWH00002B/500

9 789387 057562